Creative Rationality and Innovation

Smart Innovation Set

coordinated by
Dimitri Uzunidis

Volume 14

Creative Rationality and Innovation

Joëlle Forest

iSTE

WILEY

First published 2017 in Great Britain and the United States by ISTE Ltd and John Wiley & Sons, Inc.

ISTE Ltd
27-37 St George's Road
London SW19 4EU
UK

www.iste.co.uk

John Wiley & Sons, Inc.
111 River Street
Hoboken, NJ 07030
USA

www.wiley.com

Library of Congress Control Number: 2017953282

British Library Cataloguing-in-Publication Data
A CIP record for this book is available from the British Library
ISBN 978-1-78630-146-8

Contents

Foreword

This book is a fascinating reflection on the concept of innovation. Initially, it gives a very vivid account of what differentiates innovation from Research and Development by addressing both economic and historical aspects. The contribution of the history of innovation models, initiated by Schumpeter's models of innovation and subsequently studied in depth and revisited by various authors, is highly useful for engaging in reflection on richer and more robust models. This makes it possible to introduce the concept of creative rationality, an indispensable factor at the very core of the emergence of innovation.

Although innovation is the driving force of economic development and value creation, this concept, which is constantly bandied about as the solution for all our evils, sometimes conceals a lack of understanding of what it stands for by those who refer to it. Contrary to what our decision makers would have us believe, technological inventions, basic research and scientific discoveries arising thereof neither naturally lead to innovation nor to the hoped-for economic benefits. Innovation aims to develop a product or service that meets user needs and contributes to the dynamism of the economy. Basic research aims to understand the world around us. It must be borne in mind that these objectives are radically different from innovation objectives. As such they need not merge into one another, let alone be subjected to one another. Let us accept that basic research and innovation correspond to different objectives and obey their own dynamics. Although the innovation process does not come about directly from research endeavors, it often benefits from scientific advances.

How can we foster innovation, and ensure that we are really responding to market needs that are often, but not always, expressed[1]? A process that takes the end user into account and places him at the center of the design process makes perfect sense. Design is an organized process that leads to the making of a product suited to the end user's need. Design is the guiding principle of the innovation process by its ability to define the challenges that the innovative product must meet. This process is different from the R&D process as it implements a process driven by the end purpose of the object, but at the same time, creativity is not absent in this approach. It is in fact a key element that opens the scope of possibilities to finally retain only the most suitable one. In her book, Joëlle Forest compels us to reflect upon all these aspects and proposes a methodology to engage this reflection.

The notion of creative rationality thus comes up: far from opposing one another, creativity and rationalism enrich one other. The role of creativity lies in the opening up and exploration of all the possibilities, it must be followed by a rational stage that will make it possible to classify, select and develop the best solutions using experimentation: choosing whatever works best. Through a number of

1 A market need that is expressed corresponds to an obvious fact that meeting it will be a commercial success (e.g. a vaccine against AIDS). A non-expressed need corresponds to a need that the market does not necessarily perceive before the object comes into existence but which will become a commercial success in spite of everything (e.g. iPad).

examples that convincingly illustrate the author's thinking, we will be able to break down the innovation process and consider how creative rationality unfolds during such a process. The analysis here is very rich, and it is based on a reflection that relies on references while coming out with an original approach.

This book also paves the way for advocacy for the introduction of innovation as a course in our education system. More generally speaking, Joëlle Forest argues in favor of education that must give full scope to creativity: a less normative and more open education. Innovation can be learned only with the desire to stray off the beaten track and to implement new processes and refrain from always using the old tried and tested methods. How can we not subscribe to such a vision and how can we not support an approach that opens up exciting perspectives for our youth?

Didier ROUX
Director, Research and Innovation
of the Saint-Gobain Group
Member of the Academy of Science
and the Academy of Technology
Professor at the Collège de France (Liliane Bettencourt
Technological Innovation Chair 2016–2017)

Introduction

It's June 2017. Severely weakened by an unprecedented economic crisis, the Martian government is struck with full force by the Terra epidemic, the origin of which is a bacteria from the Earth of the genus *Bacillus* introduced by the NASA Curiosity robot that had come to explore their planet. There is discontent among the population, and the Martian government is afraid that there may be a social revolt if it does not immediately come up with concrete solutions. How can they cope with such an unprecedented situation? The Martian government decides to send Professor MacGyver junior to carry out a benchmark study of extra-Martian practices and to go and see how earthlings solve their problems.

Shortly after landing on Earth, in France more specifically, MacGyver junior noticed that the French have a miracle solution; it is called innovation. In fact, everything happens as if life on Earth were governed by innovation.

Several clues made him reach this conclusion. The importance given to innovation in France can, at the outset, be assessed based on the popularity of the concept of innovation in France. To do this, on September 23, 2016, he conducted a search on a search engine that the French call Google to see how many times the term innovation comes up in the French language. The result of this exercise is highly enlightening: 407,000,000 hits. In comparison, the terms growth and unemployment, which seem to refer to highly sensitive issues for the French government and the French people, only appear 53,200,000 and 2,860,000 times, respectively. He specifies that though it is obvious that not all of these results are pertinent, the frequency of their occurrence is nevertheless illustrative of the interest the French seem to attach to the issue of innovation.

MacGyver emphasizes that the importance given to innovation in France is closely linked to the virtues that people associate with it. Today innovation is clearly earmarked as:

– A vital necessity for companies to adapt to the requirements of their markets and to anticipate these requirements. He was privy to a narration of the story of a company called Nokia, which was the leader of the mobile telephony sector. As the said company had not thought of betting on touchscreen technology, it was overtaken by new entrants. Adapting to market requirements appears to be all the more necessary when we consider the market share of the turnovers made with newly marketed products. As an example, the turnover made with products marketed for less than two years on the small electrical household appliances market in France rose from 17% to 58% between 1997 and 2007.

– The means to revive the economy and get back to the path leading to prosperity. As highlighted by the Nobel laureate for economics, E. Phelphs, in 2006, speaking about the unprecedented economic development that took place from 1820 to 1870 in Great Britain, the Americas and subsequently in France and Germany [PHE 13], this development cannot be attributed solely to the increase in capital stock, productivity or even to the expansion of trade. This unprecedented economic development was due to innovation, and it is precisely because innovation has been going through a crisis since the 1960s that we are witness to the decline of Western countries.

– A way of meeting the major challenges of the contemporary world, as we can read in the introduction of the reference text presenting the ambitions of the European initiative "the union of innovation":

> "*At a time marked by restriction in public expenditure, significant demographic changes and strengthening of global competition, the competitiveness of Europe, our ability to create millions of new jobs to replace those destroyed by the crisis and, more generally, our future living standards depend on our ability to encourage innovation in products, services, business and social processes and models. This is the reason why innovation was placed at the very core of the Europe 2020 strategy. Innovation is also the best means we have at our disposal to solve the main problems faced by our society, which become increasingly pronounced by the day, be it climate change, shortage of energy or the increasing scarcity of resources, health issues or the aging of the population*" [COM 10, p. 2].

– Or even a means to give back a positive dynamic to societies lacking in progress and afraid that they will no longer be able to advance.

This way of looking at innovation is not neutral because it leads to concrete actions. In France, this representation of innovation gave rise to the creation of dedicated ministerial institutions, such as the bureau for innovation, and technology policies at the Direction générale de la compétitivité, de l'industrie et des services (DGCIS – Directorate General for competitiveness, industry and services); the creation of innovation ecosystems such as the pôles de compétitivité (competitiveness clusters) with the stated objective of stimulating the dissemination of knowledge at the root of innovation: "*The bringing together of industrial, scientific and training stakeholders from the same territory (...) constitutes in itself a source of innovation*" [CIA 07]; and this representation of innovation also led to the definition of priorities in 2009 within the framework of the National Research and Innovation Strategy; the formulation of dedicated plans such as *Une nouvelle donne pour l'innovation* (*A game changer for innovation*) in 2013; the production of a number of reports such as the Morand–Manceau report in 2009 [MOR 09], the Godet–Durance–Mousli report in 2010 [GOD 10], the Birraux, Le Déaut report in 2010 [BIR 10], the Beylat–Tambourin report in 2013 [BEY 13], or the Pisani-Ferry report [PIS 16]; and finally the creation of tax incentives such as the Research tax Credit (CIR – Crédit d'impôt recherche) which today amounts to around €6.5 billion.

The importance of innovation in society is at last perceptible in view of the number of scientific articles, reviews and books devoted to this topic, and the existence of communities of dedicated researchers such as the Innovation Research Network.

Armed with these observations, Professor MacGyver junior returns home and submits his report to the Martian government. His conclusion is just one sentence long: if we wish to get out of the crisis and defeat the Terra epidemic, we must innovate! The world is not immobile but constantly changing. We must therefore draft an innovation policy which will allow us to not only face contemporary challenges but also to invent our future. Highly enthused, the government hastens to pass on the conclusions of Professor MacGyver junior: let us innovate, everyone innovate! Yes but how do we go about it?

The reader would have understood that we have used the parable of Professor MacGyver junior as a revelatory tactic, in the photographic sense of the term, to depict the situation in which we find ourselves in France.

Over the last 15 years, financial assistance given to innovation by public authorities, estimated today at ten billion euros, has in fact doubled. This assistance is intended to serve a national ambition whose aim is to transform our old "*economy of imitation*" into an "*economy of innovation*" [AGH 04]. Despite this, we must mention that the increased number of mechanisms (we have apparently gone from around 30 assistance mechanisms in the 2000's to 62 today) [PIS 16] is not producing the expected results:

> "*A number of initiatives, quite often pertinent ones, have been taken to foster the development of innovation (...). This has resulted in a barely comprehensible accumulation and diversity of mechanisms and structures, both at the national and the regional or local level, the overall, economic, industrial and social effectiveness of which remains to be demonstrated (in terms of job creation)*" [BEY 13, p. 1].

What is worse is that the speed at which these mechanisms are being renewed could even favor "*bounty hunters*" [PIS 16, p. 33].

The question that arises is: what should be done to go beyond the stage of demanding innovation[1] and garner an effective capability to innovate?

According to Pisani-Ferry, a response to this issue would entail decreasing the number of mechanisms because it is "*difficult to believe that the State has the wherewithal to supervise a set of 62 mechanisms in a consistent manner and guarantee proper coordination with those initiated by local authorities*" [PIS 16, p. 31].

In this book, we will defend the thesis according to which innovation has to be thought about differently. From this viewpoint, the report of the national commission for evaluation of innovation policies *Quinze ans de politiques d'innovation en France (Fifteen years of innovation policies in France)*, steered by J. Pisani-Ferry, is edifying. He shows that, in 2014, 70.2% of the allocated resources were related to the growth of private R & D capabilities. As we will see, this conception of public policy is founded on the innovation model inherited from J. Schumpeter. However, we will show that if we consider innovation from its central process, namely the design process itself, we will feel compelled to consider other possible means of action. By adopting an artificialistic perspective, this book lays emphasis on creative rationality. It considers the implications of this point of view for teaching. And while we are at it, this book will dispel a number of myths surrounding innovation.

1 X. de la Porte, a French journalist and essayist even refers to the demand to innovate: "*The demand to innovate contains the idea that the way to get out of the crisis is to march ahead, march towards novelty, produce. (A linear concept of civilizations which was already inherent to the notion of progress.)*" [DEL 14].

1

Innovation: What Exactly Are We Talking About?

Contrary to what we tend to think, innovation is neither a fashionable preoccupation nor an embodiment of capitalism. The history of techniques makes us understand that innovation is an integral part of humanity [JAC 11]. The philosopher M. Puech confirms this viewpoint. According to him, technique has been by man's side right from the origin of humankind, it has been at the core of humanization, whereas science has only been by our side since relatively recently. This joint evolution of homo sapiens with technique even makes him allude to the term homo sapiens technologicus [PUE 06]. However, the issue of what precisely the concept of innovation covers arises. We will demonstrate that it is the ability of innovation to create value that allows us to distinguish it from invention. This value creation can be seen in innovations which take on various forms and have an impact which is also varied in nature.

1.1. Some key distinctions

1.1.1. *Distinguishing innovation from discovery and invention*

The third edition of the Oslo Manual [OEC 05] defines innovation as the implementation of a product (a good or a service) or a new or distinctly enhanced process, a new marketing method or a new organizational method in corporate practices, the organization of the workplace or external relations.

From this point of view, innovation is defined according to the novelty criterion and the Oslo Manual specifies that it could refer to a novelty for the company, a novelty for the market or a novelty for the whole world if the concerned firm is the first to launch it in all the markets and in all business sectors.

As the novelty criterion also applies to the concepts of discovery and invention, we feel that it is important to demarcate innovation from discovery and invention.

Discovery is the act of finding something that is hidden. Such as the discovery of the double helix structure of DNA by J. Watson and F. Crick in 1953 for which they were awarded the Nobel Prize for medicine in 1962. Also, the discovery of the giant magnetoresistance principle by the physicists A. Fert and P. Gruenberg in 1988 for which they received the Nobel Prize for physics in 2007. Thus, and as pointed out by the philosopher of science M. Serres[1], only things which were already in existence but remained unknown, things which had not yet been seen, can be discovered. For example, the development of a technique to manufacture microscope lenses of a quality and power unknown until then elsewhere in the scientific world at that

1 Interview can be accessed online: http://fresques.ina.fr/jalons/fiche-media/InaEdu04626/la-decouverte-scientifique-selon-michel-serres.html

time is what led to A. van Leeuwenhoek's announcement of his discovery of the human sperm at the Royal Society of London in 1678. More recently, the HARPS spectrograph installed on a telescope of the ESO (European Southern Observatory) in Chile and measurements carried out between 2000 and 2014 on ESO's telescopes made it possible to discover that Proxima Centauri, the nearest star to the Sun, had a planet[2]. Published, in August 2016, in Nature, these research works show that this exoplanet, baptized Proxima b, is rocky and has a mass that is comparable to Earth's mass.

Discoveries can materialize through inventions. There are numerous examples of "discovery followed by invention" pairs. By way of example, we can cite electromagnetism and electricity networks, thermodynamics and spark-ignition engines. Contrary to discovery, invention and innovation are artificial. They make a hitherto unseen reality happen. Considered as such, discovery comes within the scope of research, whereas invention and innovation come within the scope of engineering. Discovery belongs to the past, whereas invention and innovation belong to the future.

1.1.2. *What is the distinction between invention and innovation founded upon?*

With invention, we move away from the realm of natural and formal sciences and draw closer to the technical domain so much so that invention is often mistaken for innovation.

In fact, in both cases, we end up with an outcome that seems like something new and the structure of their creation process is identical (each one has a cost associated with it, implying allocation of specific means, each one needs time to emerge, proceeding by trial and error, going back and forth,

2 The relation between scientific discovery and development of technical objects may be noted.

etc.). However, it is necessary to distinguish invention and innovation, because not every invention systematically translates into an innovation[3].

At this juncture, we can cite the example of the invention of the digester by D. Papin in 1679. D. Papin's digester is a kind of cooking pot in which water changes to steam, and then pressure and temperature build up until they attain the level of pressure fixed by the safety valve. Fitted on the cover, the steam-release valve keeps the digester from exploding.

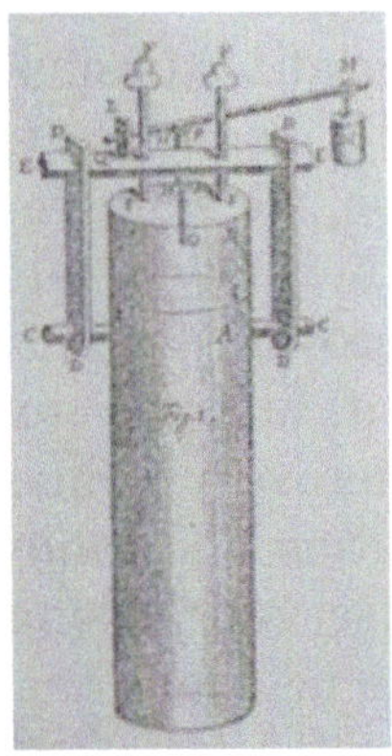

Figure 1.1. *Papin's digester (source: http://fr.wikipedia.org/wiki/Denis_Papin)*

In his book titled "*La manière d'amolir les os et de faire cuire toutes sortes de viandes en peu de temps et à fort peu de frais*", after describing the digester, D. Papin recounts his discoveries and experiments on cooking, in particular of meats, fish, vegetables, etc. And because people object, he says, to new inventions claiming that the expenses incurred will be greater than the profits they could generate, he devotes his ninth chapter to "Price calculation" and demonstrates the profitability of such a machine. But less inclined to trade, he did not make a commercial application

3 We will discuss this point at greater length in the course of Chapter 2.

of his invention. It was in 1953 that SEB innovated from D. Papin's concept and launched the SEB Super-Cocotte (pressure cooker).

Over the past few decades, other French and foreign companies have produced pressure cookers based on the same principle of cooking through an increase of pressure and temperature. The advantages of this very first French pressure cooker, manufactured by the stamping process, are to be found in its robustness and lightness, its aluminum body which is a good conductor of heat, the fact that it is easy to clean and its price, which is two times lower than the average market price.

Value creation is all the more apparent when we see that the Super-Cocotte perfectly met requirements at the time. Barely three years after the end of ration cards, households were keen to give proper nourishment to their families within a tight budget. Although beef was the main component of daily food at that time, it had the disadvantage of requiring cooking over a long period of time, especially for the least expensive pieces. In other words, the money saved by buying cheaper cuts of meat was spent on gas to cook it. This economic sales pitch was therefore widely projected at the time of its launch.

Figure 1.2. *Super-cocotte SEB, 1953 (source: http://www.seb.fr/marque/historique.html)*

The preceding example allows us to note that the main characteristic of innovation is to meet a requirement that can be stated or latent. The notion of need is important to the extent that it underscores the point that innovation does not necessarily respond to an explicit demand from a consumer, but can sometimes be forced on him/her, which explains the idea of a latent need. By doing this, we move away from the definition set out in the report of the *Office Parlementaire d'Evaluation des Choix Scientifiques et Techniques* (The Parliamentary Office for Scientific and Technological Assessment) entitled Innovation Facing Fears and Risks, according to which "*Innovation is the art of integrating the best state of knowledge at a given point in time in a product or service in order to meet a need expressed by citizens or by society*" [BIR 12, p. 15][4]. We align ourselves with J. Schumpeter, who already emphasized in Business Cycles that innovation does not necessarily come in response to an expressed need:

> "*Railroads have not emerged because any consumers took the initiative in displaying an effective demand for their service in preference to the services of mail coaches. Nor did the consumers display any such initiative wish to have electric lamps or rayon stockings, or to travel by motorcar or airplane, or to listen to radios, or to chew gum.*" [SCH 39, pp. 67–68].

Concretely speaking, the idea of need refers to the ability of innovation to create consumer value. What does this mean exactly? Consumer value is assessed in relation to the perceived benefits (promise of an innovative offer) by

---

4 As we will see further ahead, disruptive innovations, or innovations with regard to the identity of products, do not necessarily originate from a demand.

the said innovation[5]. However, the consumer value is "reduced" by:

– the perceived cost of innovation, which includes the purchase and maintenance costs related to the acquisition of the said innovation, and also the time and effort required to appropriate the said innovation. Is this innovation easy to use or does it require a training period? For instance, changing one's smartphone is likely to entail certain costs to become familiar with a new interface;

– the risks that the user associates with the use of the innovation. Microwaves, just like cell phones, are the subject of major controversies even today with regard to their health hazards. In January 2010, the production of *Google Glass* was suspended, owing to its sale price (1,499 dollars) and also due to the hostility of people who were uncomfortable with the idea of being filmed without their knowledge. Similarly, users are bound to ponder over interoperability hazards.

An invention whose value creation has not been demonstrated will thus never reach the innovation stage. This explains quite precisely the fact that very few inventions presented at the Lépine competition get into the public domain.

We must, however, point out that value is a subjective notion. What might seem "useful" to a person might not necessarily be so for others. Let us take the example of skis. A mother would prefer a ski that comes off easily in case of a fall for the sake of her children's safety and in order to avoid risks of a fracture. On the contrary, an extreme skiing enthusiast would much rather go for a tight fit which would come off less easily. Likewise, although today we find cell phones whose prices range from 10 euros to more than 700,

5 We must specify that value creation for the user should not disregard the issue of economic viability necessary for the deployment of the said innovation. We will discuss this point subsequently.

it is quite simply because we don't all give the same importance to having a multi-functional phone, to video quality, to the speed offered, etc.

Moreover, value is a social construct as we can see in the case of toilets. Because the way we equate cleanliness is not the same in each country, we do not purchase the same WCs. In countries where toilet paper is not used (because it is believed that paper does not clean thoroughly), people wash with their hand. In these countries, adding small taps located below the toilet contributes to creating value. Similarly, in Japan, we find features in toilets which are likely to be considered rather pointless in other countries: a heated seat, an air dryer, adjustment of the position of the water nozzles, etc.

We would like to lastly specify that value creation implies that the designed object is not only useful but also usable. This is the reason why chindogus, unusual or weird tools, cannot be considered as innovations.

1.2. Typology of innovations based on their purpose

Value creation can take place both small enterprises and leading multinational companies. It is not restricted to certain business sectors. We can innovate in the automobile industry, the habitat segment, the food-processing industry as well in case of products that are as unremarkable as toilet paper.

On the contrary, innovation is protean in nature. In fact, innovation can relate to:

– Products (such as the Bluegard® autonomy bracelet; endowed with a geolocation system and a 24/7 monitoring system, connected to a platform; devised for Alzheimer's patients), or a service feature, such as home help for the elderly.

– Processes such as hydroponic agriculture without soil (wherein soil is replaced by a sterile substrate, such as clay pebble substrates or rock wool substrates) or second generation biofuel production from agricultural residues.

– Marketing. In this case, it can be based on a new distribution method. For example, to go with the consumer's wish to opt for local and neighborhood channels, to get back to direct purchase from the producer; it is also egged on by the craze for bulk offers which makes it possible for Auchan to throw open its doors to 13 million single persons and 2.5 million students in France who happen to be consumers with moderate requirements and therefore do not want to be forced any longer to buy quantities imposed on them and who thus finally discover the feeling of having control over what they consume [DEL 15]. Marketing innovation can also be based on a new pricing principle (low-cost-based strategy) or on a new packaging policy (Pom'Potes compotes from Materne in their flask format meet the consumers' current trend to be on-the-go and the need to have easy-to-carry balanced children's snacks).

– Organizational. As an illustration, we can quote the 80/20 principle initiated in the 1930s by W. McKnight, the then CEO of 3M, which aimed at allowing employees who wished to do so to devote 20% of their time (i.e. 1 day per week) to working on projects of their choice. This approach, which gave birth to post-it in 1974, has now been adopted by Google through their famous 70-20-10 law.

– Social, i.e. innovations which "simultaneously meet social needs and create new social relations or new collaborations" [MUR 10]. For illustrative purposes, it is fair trade that contributes to creating the conditions of international solidarity between the North and the South. Yet, another thought that comes up at this juncture relates to micro credit initiated in the 1970s in Bangladesh by Professor M. Yunus, winner of the 2006 Nobel Peace Prize, who wished to enable the poor women of his country to indulge in modest economic activities.

Considering innovation from the forms it can take allows us to escape the confines of a restrictive definition of innovation with reference to the existence of a market. As emphasized by T. Ménissier:

> "*social innovation could be initiated by citizens, and emerge in the social space of political demands; this is clearly testified by the great social innovations of the 19th century favorable to solidarity: graduated tax, right-to-work, social security*" [MÉN 11, pp. 12–13].

The consideration of the protean nature of innovation also validates the idea according to which the thing that denotes the innovative character or otherwise of an artifact does not lie in its technicality but in its ability to create value for its intended user.

1.3. Typology of innovations based on their scale

Even though all innovations have the characteristic of creating value for their intended end user in common, not all innovations have the same impact on society.

In fact, most innovations are incremental in nature. The main characteristics of this type of innovation are their frequency and autonomy, which implies that they do not make any in-depth changes to the existing established firms, no more than the product's production and marketing mechanism. The shift from glass Coca-Cola bottles to Coca-Cola cans does not entail any change in the industrial structure. Likewise, in the telecommunications sector, the move from 2G to 3G and subsequently 4G which improves the scope of digital data transmission would come under the realm of incremental innovation. This type of innovation seems relatively easy to implement and control, the inherent level of risk is generally limited with regard to technical and marketing plans, and its financial impact is also less significant.

On the contrary, radical or disruptive innovation, though rare in occurrence, has major consequences on the entire production and marketing system and is a cause for considerable difficulties for established firms. At this juncture, we can, by way of illustration, cite the example of Kodak not wanting to go digital, which was threatening its film-based business model, and the subsequent fall of its market share of the French film photography market from 76% to 6% between 2001 and 2005; Kodak failed in making the right strategic choices. This strategic failure was the direct cause of Kodak's decades-long decline which led to its bankruptcy in 2012. Another example that comes to mind is that of the French Internet access provider industry following the launch of Triple Play (Internet, telephone and television) by *Free* in 2003 which enabled Free to become the second largest French operator in 2005. By modifying the reference offer, Free forced its competitors to reposition themselves and undertake mergers; this sounded the death knell for established providers (Alice).

To assert that innovation started off being radical and that the contemporary era would appear to be marked by the stamp of incremental innovations as J-P. Boutinet [BOU 12] suggests is misleading[6]. In reality, these two types of innovations have always coexisted and correspond to

6 It may be noted that C. Christensen prefers to speak of sustaining innovation and disruptive innovation. The former is done in conformity with the company's business model, irrespective of whether it is radical or otherwise. The latter is done in disruption of the company's business model, once again irrespective of whether it is radical or otherwise.

differentiated innovation strategies. Let us take the example of Saint-Gobain, the global leader in the housing sector. Saint-Gobain is a group backed by 350 years of history. It all began with the creation of the Royal Mirror-Glass Factory in Paris in 1665. During the past 350 years, Saint-Gobain continued to constantly innovate in its processes and its products. However, there is one thing that can be clearly observed: Saint-Gobain's innovation policy is not focused on disruptive innovation, insofar as its corporate culture is concerned, as this is too risky. According to D. Roux, a member of the academy of science in France and the group's vice president of R&D and innovation, this is what precisely differentiates Saint-Gobain and Corning. This also explains the fact that for several years now, the Research & Development department and the Marketing department have been working together at all levels at Saint-Gobain. This kind of organization enables it to better fine-tune its response to market expectations, anticipate future needs and, while doing so, shorten development time.

We must, however, point out that it is true that since the present era is marked by a race against time and rising competition, we tend to craft innovations in a hurry and favor incremental innovations. We have shown, for instance, how "emergency urban planning" makes us opt for widely accepted principles of solution without taking the time to question the models on which they are based[7]. As pointed out by E. Phelps, such a situation would not be as serious had the dictatorship of short-term imperatives not translated into a search for short-term improvements that generate less and less income and create fewer and fewer jobs in Western economies.

---

7 For this very fact, emergency urban planning is not really favorable to the broadening of the scope of opportunities. By bowing to these exigencies, it comes under the realm of reproduction rather than that of innovation.

1.4. Reasons for innovation

The genesis of innovations arises from various motives:

– Innovation is a means of adaptation to the changes in the environment. If we think of the competitive, technological, institutional environment, etc., it can never be permanent as it evolves constantly. These evolutions act as real incentives to innovate. We can cite the example of the removal of import quotas for Chinese textiles which came into force on January 1, 2005 in Europe. This move gave rise to more than a 50% hike in certain textile product categories. The removal of these quotas forced a number of French and European companies to reposition themselves on strong value-added products (such as smart textiles). We can also cite the issue of scarcity of certain raw materials. The coal carbonization system was, for instance, invented to solve the problems created by the scarcity of wood in England. Similarly, to manufacture soda ash, we used to import a plant called salicornia from Spain. The wars during the revolution followed by those during the Empire put a block to exchanges between France and Spain, but soda ash was indispensable for soap and glass manufacturers. It was then that the choice was made to use a process invented before the revolution, which made it possible to manufacture soda ash from sulfuric acid and sea salt.

– Innovation is also a means to adapt to the introduction of new standards. The 2002 decision of the Health Nutrition National Plan to reduce salt consumption by 20% over a period of 5 years in order to alleviate hypertension problems and cardiovascular hazards impelled the stakeholders of the food-processing industry to innovate. Similarly, the circulation of the sustainable development guidelines in the wake of the 1987 Brundtland report paved the way for the introduction of new standards. The latter are, for instance, at the root of:

– The development of new materials in construction (insulation using materials of plant origin: wood, cellulose, hemp, cotton, linen, etc.);

– The increasing use of natural fibers in the automobile industry (which replace synthetic or glass fibers);

– The increasing use of composites in the manufacturing of aircrafts. Airbus A380 is, for instance, the first commercial aircraft structure is made with 25% of composite materials. Out of these 25%, 22% of the structure is made with carbon, glass or quartz fiber composites and 3% with GLARE (an aluminum–glass fiber laminate, used for the first time in a civil airplane). As they are lighter than aluminum, composite materials decrease the aircraft's unladen weight. For instance, making the center-wing section with carbon increased the weight of the aircraft by one and a half tons. The decrease in weight ends up reducing fuel consumption and therefore leads to a decrease in operating costs and lowering of emissions released in the atmosphere;

– And even the fabrication of sustainable cities, especially after the Aalborg conference [FOR 15].

– Innovation provides a means for coping with the saturation of markets: a family of four owns around 3,000 objects, whereas it didn't have more than 200 objects 100 years ago! This situation has arisen due to the fact that goods which were earlier considered as luxury goods have now become everyday consumer goods (cars, TV, personal computers and smartphones), in particular owing to the fall in the sale prices of certain products because of productivity gains. The only way in which manufacturers who sell these types of products can regenerate demand is by innovating.

– Innovation finally appears as the means to meet major contemporary challenges such as the aging of the population, social justice, and the consideration of the guidelines for sustainable development to quote only a few.

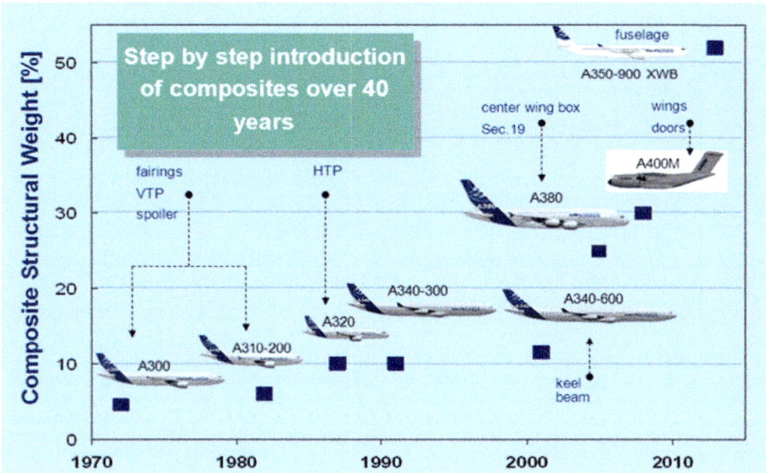

Figure 1.3. *Evolution of composites (source EADS, 2011)*

Now that the concept of innovation and the reasons for innovating have been specified, the issue of identifying possible means of action to innovate has to be addressed. In order to do this, let us examine beforehand the relationship between the innovation model and innovation management.

2

Thinking about Innovation Differently

One observation that stands out is that while there is still plenty of talk about innovation, we seem to be barely capable of innovating insofar as our way of conceiving innovation is concerned. Looking at innovation differently is not just a theoretical issue, it is also a practical issue. In fact, modeling innovation will make it possible to describe the innovation process and define a set of actions in order to obtain the desired value creation according to the invested resources. In other words, the management of innovation is based on models. We will thus see that innovation policy in France continues to be mainly based on the Schumpeter 2 model of innovation. Bearing the limitations of such a model in mind, we will briefly examine the contours of the artificialistic perspective. First, let us throw a glance at the place of innovation in society.

2.1. Innovation in society

The archeology of the concept of innovation reveals that innovation was initially viewed in a negative manner. Plato, a great philosopher of antiquity, had no hesitation in expressing his distrust towards innovation:

> *"If we amongst ourselves bestow extraordinary honors on those who are always inventing something new, introducing styles that are different from the established ones in clothes, colors and all such types of things: we can be sure, without any fear of being wrong, that there is nothing more disastrous than this for a State. In fact, without our even noticing it, this makes the youth [...] despise all things old and hold in esteem all things new [...] this is the greatest evil that can befall a State"*. [PLA 13, pp. 2679–2680]

As we can see, what Plato feared was the ability of innovation to alter mores and, in fact, the established order. This explains why innovation was at first proscribed: "*this type of interdiction manifested itself legally for the first time in 1548 when the King of England Edward VI, succeeding Henri VIII, published a proclamation explicitly forbidding innovation and threatening the detractors* with imprisonment" [GOD 14]. Such a conception of innovation continued for many centuries, as we can see from the definition given in the *Encyclopedia* of Diderot and d'Alembert: "*A novelty or significant change made in the political government of a State against the use and rules of its constitution. These types of innovation are always deformities in the political order*" (Joncourt (1751), in [HUY 13]) wherein innovation is thought of as an illness ("deformity").

A few thinkers, like J. Bentham, will gradually endeavor to show the vacuousness of such a viewpoint with respect to innovation:

> *"The word innovation is imputation of bad motives, bad designs, bad conduct and character ... Innovation means a bad change, presenting to the mind, besides the idea of a change, the proposition, either that change in general is a bad thing, or at least that the sort of change in*

> *question is a bad change. ... [But] to say all new things are bad, is as much as to say all things are bad, or, at any event, at their commencement; for of all the old things ever seen or heard of, there is not one that was not once new. Whatever is now establishment was once innovation".* (Bentham (1816) in [GOD 14])

If innovation asserted itself right from the 16th Century as a symbol of a constant breakaway from tradition, a relentless challenge thrown to the classics [MÉN 11] and heralded the advent of the era of design [FAU 01] wherein people designed the world in which they lived or imagined living in (rather than inheriting a created world which preceded them), an era marked by the engineer and no longer by the discoverer, we can nevertheless presume that the teaming up of innovation with the idea of progress largely contributed to its popularization. After all, is believing in progress not the best way to make the breakaway from tradition bearable?

In the post-war period, the "trente glorieuses" (the Glorious Thirty), characterized by growth of an exceptional length of time and magnitude, as well as by scientific and technical breakthroughs present in all economic sectors, witnessed a profusion of writings attempting to explain the role of innovation or measure its impact on economic growth and development. Many economists were sure then of "the irreversibility of the growth movement", of the constant nature of technical advancement and of innovations sustaining such a growth movement. The 1960s crisis put an end to this optimism. The return of the crisis was accompanied by a flurry of contributions which explained the reasons thereof as the end of the effects of post-war innovations and showed that the differentials of competitiveness between firms, or even between nations, were to be found in their greater or lower ability to innovate and thus penetrate or create new markets. Underlining the decisive role of innovation in coming out of the crisis and in competitiveness, a number of contemporary economists contributed to giving back credibility to the writings of J. Schumpeter.

2.2. Schumpeter's models of innovation

In "Petite histoire des modèles d'innovation" [FOR 14], we have shown how the economic analysis of innovation shifted from an economic analysis of innovation as an outcome to that of innovation as a process.

In fact, in his book *The Theory of Economic Development*, J. Schumpeter asserts that innovation[1] is the driving force of the dynamics of capitalism [SCH 12]. He places at the very core of economic change the entrepreneur, who, according to him, is the one who implements innovation and thereby destabilizes established routines and positions.

The Austrian economist emphasizes the distinction that exists between the role of the inventor and that of the entrepreneur. The former produces inventions. The latter grabs opportunities to innovate from within a given stock of inventions, in other words scientific and technical novelties, with a view to deriving profits. Therefore, for J. Schumpeter, the dynamic of invention is not an issue coming within the realm of economics: "*the invention [...] was not an external factor of the business situation of its time; it was, indeed, no factor at all*" [SCH 39, p. 15]. According to him, defining

1 We may recall that by innovation J. Schumpeter meant "*the introduction of new commodities which may even serve as the standard case. Technological change in the production of commodities already in use, the opening up of new markets or of new sources of supply, Taylorization of work, improved handling of material, the setting up of new business organizations such as department stores – in short, any 'doing things differently' in the realm of economic life – all these are instances of what we shall refer to by the term Innovation*". [SCH 39, p. 80]

innovation by invention would mean putting the focus on an element without relevance to economic analysis: "*invention [...] produces of itself no economically relevant effect at all. [...] defining innovation by invention would, therefore, [...] mean stressing an element without importance to economic analysis*" [SCH 39, p. 81].

On the contrary, innovation is a matter coming within the scope of economic analysis to the extent that it takes part in the dynamics of the economic system. "*As soon as it is divorced from invention, innovation is readily seen to be a distinct internal factor of change. It is an internal factor because the turning of existing factors of production to new uses is a purely economic process and, in capitalist society, purely a matter of business behavior*" [SCH 39, p. 82].

This vision of a strict separation between the scientific and technical domain and the economic domain is not specific to J. Schumpeter. This even seems to have been the dominant mode of representation with economists from the late 19th Century to mid-20th Century. It is at the root of the black box model[2] (Figure 2.1).

Figure 2.1. *Black box model*

2 The "black box" concept is borrowed from the title of a book written by S. Kline and N. Rosenberg.

At the input are inventions which, under the action of a specific actor, the entrepreneur[3], are transformed into innovation.

Innovation is defined as the outcome of a choice. It is the product of the economic act.

The transformation function is included in the black box and because of this, it is not specified by the model. In fact, "standard" economic theories restrict themselves to considering and analyzing the consequences of technical progress on economic dynamics. However, studying the consequences of a technical change without understanding its origin or its content, as done by traditional economic analysis, not only seriously restricts the scope of reflection but also refrains from accessing the means of a policy right at the outset, whether economic or technological in nature. The main criticism leveled against such theories pertains precisely to this point. In fact, can we consider a model which considers the repercussions of a given phenomenon, innovation in the present case, on economic dynamics without having even considered its genesis robust enough?

2.3. From innovation as an outcome to the analysis of innovation as a process

A little before the second half of this century, the approach of innovation as an outcome was gradually substituted with the approach of innovation as a process. Although the change of the industrial and competitive environment of companies seems to have largely contributed to the adoption of a new approach of innovation, we must nevertheless refrain from making it the sole explanatory

3 Let us note that J. Schumpeter turned down the predominant paradigm of entrepreneurship inherited from neoclassical growth theory as management of the firm and replaced it with an alternative one: the entrepreneur as the innovator.

factor. The black box model also seems to have been challenged by the reality check, especially with the acceleration of research and development investments: "*Can we find any better proof of the force imposed upon observers by these changes than the case of J. Schumpeter himself, who, barely three years after completing* Business Cycles, *noted in* Capitalism, Socialism and Democracy, *not the stop of economic progress but the 'dusk' of the function of its traditional promoter?*" [MAU 68, p. 14].

In 1942, i.e. eight years before his death, J. Schumpeter had already become aware of the rise of in-house corporate research and development (R&D) in large firms in the 20th Century. He asserted with great insistence that technical change owes its presence less and less to an individual entrepreneur but instead is the fruit of organized work conducted in research and development departments. The consequence that the Austrian economist drew concerning this was two-fold:

– invention and innovation are henceforth commonplace activities of most firms;

– the theory of the creative function of the company succeeded that of the genius entrepreneur.

We thus see a radical change of perspective with respect to the black box innovation model. Indeed, admitting innovation as a commonplace activity of firms challenges the role of the Schumpeterian genius entrepreneur, which in turn leads to challenging the random nature attributed to innovation and makes it possible to reflect on the manner in which the innovation process takes place and can be improved. Moreover, the shift made from the economic analysis of innovation as an outcome to that of innovation as a process paved the way to the linear and hierarchical model of innovation (or Schumpeter model 2).

2.4. Contours of the linear and hierarchical model of innovation

The starting point of the linear and hierarchical model of innovation is R&D[4]. The interest given to R&D was stimulated by an analytical observation, namely the difficulties encountered by some economic theories of growth. Unlike standard economic analysis, considering R&D as a factor for the genesis of innovation consists in fact in endogenizing technical progress. However, the attention on R&D was also linked to empirical observations, namely:

– the sustained growth of R&D expenditure;

– the creation of the first companies originating from the application of new scientific discoveries such as Siemens or Bayer right from the late 19th Century;

– the history of reputed innovations such as radiography (following the discovery of X-rays at the end of 1895 by the German physicist W. Röntgen) or nylon stockings (sold from 1940 by the company DuPont de Nemours following the discovery of the formula for nylon by W. Hume Carothers in 1935).

This model is said to be linear insofar as the innovation process is represented as a succession of stages which act as so many compulsory checkpoints that have to be gone through. At the beginning of the innovation process lies research activity, which is followed by development, production and then marketing of a new product.

This model is said to be hierarchical because it is presumed that the outputs of a considered stage constitute the inputs of the following stage. In other words, each stage must be completed before moving onto the following stage.

4 The linear and hierarchical model is also called the big science model in the literature by some authors.

With this in mind, tasks are carried out in a sequential manner and cannot be conducted in parallel (anteriority constraint). This mode of functioning makes use of a partitioned organization and specialization of individuals or activities.

Described using a linear and hierarchical model, the innovation process appears thus as a sequential process following a perfectly predictable sequence (Figure 2.2).

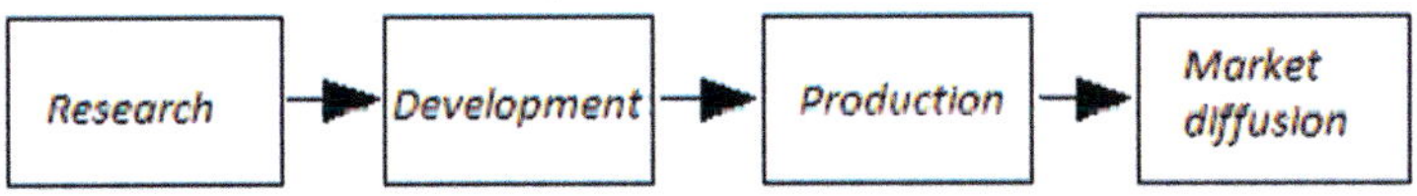

Figure 2.2. *The linear and hierarchical model of innovation (adapted from Kline and Rosenberg in [FOR 14])*

2.5. A fertile ground for the creation of the linear and hierarchical model of innovation

2.5.1. *The institutionalization of science*

Academies are the first institutions of science. In Europe, the first and foremost science academies came up in Italy: the Lincean Academy (Accademia Nazionale dei Lincei), founded in Rome in 1603 and the Academy of Experiment (Accademia del Cimento), created in 1657. This movement took a decisive turn with the creation of the Royal Society in London in 1662 and then with the creation of the Royal Academy of Sciences (Académie royale des sciences) in 1666 in France.

The creation of academies meets multiple challenges:

– Before the Renaissance, science remained restricted to a few initiated persons and the practice of scholars was to retain their discoveries to themselves. On the contrary, academies solicited scientists to compare their discoveries. They held sessions during which research was presented and published the minutes of these sessions in the form of science journals, contributing by doing so to opening access to science.

– As P. Papon [PAP 01] points out, the emergence of these academies materializes an idea put forth by F. Bacon in the *Nouvelle Atlantide*, published in 1627: science must be organized and systematically applied to the industry. This is particularly true of the creation of the Royal Academy of Sciences in France by Colbert, as the latter had understood that scientific progress could lead to technical progress capable of increasing the power of France and the glory of the king.

Having become a full-fledged institution at the end of the 19th Century and in the early part of the 20th Century, science got transformed into a new religion, or into what F. Le Dantec qualified as scientism:

> *"I believe in the future of Science: I believe that Science and Science alone will solve all the questions which have a meaning. I believe that it will penetrate up to the arcana of our sentimental life and that it will explain to me even the origin and the structure of the anti-scientific hereditary mysticism which coexists within me along with the most absolute scientism. But I am also convinced that men wonder about many questions which have no meaning at all.* As far as these *questions are concerned, Science will show their absurdity by not responding to them, which in itself will prove that they do not have an answer".* [LED 12, p. 55]

Scientism will thus be similar to a sort of non-critical and ultra-enthusiastic attitude towards science, less inclined to acknowledge its limitations and its potential hazards. Science is capable of everything, it develops knowledge, it can solve all problems and it ensures maximum happiness to mankind. The reality seems to validate this ideology. The development of knowledge in physics (and in electronics in particular) ushered in the era of new information and communication technology, contributing to currently changing concepts of time and space and revealing the crucial nature of their role in experiencing temporality, spatiality and mobility. Closer to home, the research works of the French woman E. Charpentier and the American woman J. Doudna are revolutionizing the field of genetics with the Crispr-Cas9 system. The Crispr-Cas9 system offers the possibility of curing people suffering from genetic diseases by targeting the DNA of a given gene, severing it and replacing it with a variant, just like when we replace a Lego piece with another, making it like child play.

The primacy given to science in progress will appear all the more obvious[5] when the technique is unthought of.

2.5.2. *The lack of technical thought*

If the status granted to science appears as a fertile ground for the development of an innovation model centered on research, we must, however, emphasize the fact that the lack of a veritable technical thought contributes equally to the diffusion of this model. In fact, Greek thought is not "technological" in the contemporary meaning of the term. Although Greek thought secularizes technique (removing it from the realms of magic or divine interventions invoked until then to explain technical prowess), it is an age of *"technical stagnation"*, without a veritable *"technical thought"* despite the quality of its *"intellectual tooling"* [LAM 06, p. 28]. Thus, at the price of a philosophic tradition

5 "On fundamental research depends the progress in our ability to fashion the world that surrounds us, free ourselves from the shackles of evolution and throw open our windows to the universe and the future" [LEH 13].

that we are still heirs of, the technical object does not seem like a worthwhile object of knowledge. However, as F. Sigaut pointed out "*science cannot exist if its object is not legitimized by society*" [SIG 94, p. 59]. Although the Encyclopedists' project paved the way for descriptive technology, J. Bigelow was the one who systematized the use of the word technology in his book *Elements of technology* [BIG 29]. As a botanist and professor at the Rumford chair of Harvard devoted to "the application of science to useful arts", he defended the viewpoint of a science that is entirely mobilized by its technical applications. This is a viewpoint that we find very strongly expressed in the first half of the 20th Century, evidenced by the comments of F. Le Dantec, as previously indicated, or the slogan of the poster of the 1933 Universal exhibition of Chicago: "*Science discovers, industry applies, man adapts*".

Reduced to applications, technique continues to be rejected outside of the science perimeter[6]. However, this representation of technology interpreted as an effective practice founded on the application of science has also constituted a substrate conducive to the emergence of the linear and hierarchical model of innovation.

2.6. Impact of the model with respect to the definition of research and innovation policies

Right from the mid-1950s, the linear and hierarchical model helped define the orientations of the initial research and innovation policies. Precisely because it presents research as being the factor at the very root of innovation, it led to a rise in research expenditure. In fact, during the 1960s, R&D expenditures saw a three-fold increase and the

6 The depreciation of technique to the rank of servile activity, apart from culture, will make technique something unthought of, a label associated with objects which, although having an acknowledged existence, have an existence that is not questioned.

share of research in the GDP doubled. Of the total research expenditure in France, 70% was funded by the State. The 1970s witnessed a slowdown. The State decreased its effort but companies increased theirs considerably, from 29% in 1967 to 44% in 1979. The 1980s witnessed the return of the growth of R&D activities with a 3.7-fold increase between 1979 and 1991. The GERD[7]/GDP ratio rose from 1.8% to 2.4%, indicating the considerable effort made both by the State and companies in favor of research [MUS 98, p. 20].

Even today, the linear and hierarchical model continues to be the dominant model of innovation in France[8]. This is why France is placed third in the ranking of OECD countries which give support to companies to innovate (Figure 2.3)[9].

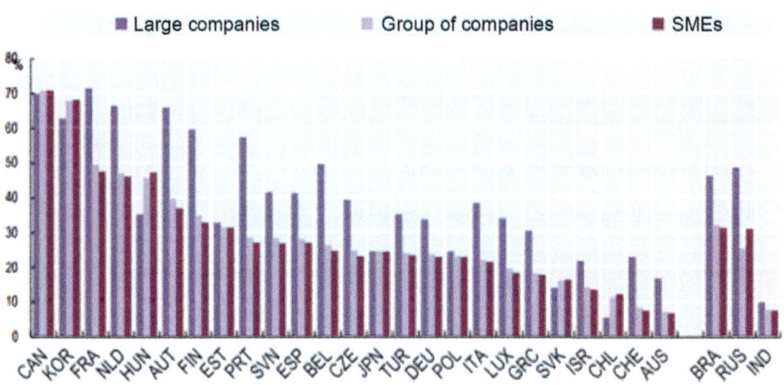

Figure 2.3. *Share of innovative companies receiving aids to innovate (source: STI Scoreboard, OECD 2015)*

7 GERD: Gross domestic expenditure on R&D.

8 It may be noted that this model is also illustrated in the objective stated within the framework of the Lisbon Strategy, impelled by the European Council in March 2000, to invest 3% of the community GDP in the area of research and development.

9 It may be noted that, in France, large companies receive more support than SMEs.

To be precise, and as indicated by the report of the national commission for evaluation of innovation policies steered by J. Pisani-Ferry [PIS 16], it is estimated that, in 2014, 70% of the allocated resources were directed to the growth of private R&D capabilities (Figure 2.4). The remaining 30% of the aid was allocated to the increase in the economic benefits of public research, the development of collaborative projects between various players, the promotion of innovative entrepreneurship and the development of innovative companies.

	Total (%)	
Family of objectives	2000	2014
1) To increase private R&D capabilities	69	70.2
2) To increase the economic benefits of public research	1	2.6
3) To develop collaborative projects between various players, networks	8	7.2
4) To promote innovative entrepreneurship	2	3.6
5) To support the growth of innovative companies	19	16.4
Total	100	100

(1) Excluding aids allocated to the defense sector

Source: CNEPI 2015

Figure 2.4. *Evolution of the breakdown of State aid in favor of innovation by family of objectives (source: [PIS 16, p. 60])*

2.7. Limitation of the linear and hierarchical model

2.7.1. *Too much importance given to R&D*

Although the linear and hierarchical model of innovation is at the root of research and innovation policies, it should however be noted that although France is well placed insofar as the indicators of averages are concerned, it appears less efficient insofar as the indicators of outcomes are concerned. In fact, *"the situation of France in the European or global landscape of innovation has remained stable for several years (11th rank in Europe, 16th rank globally) whereas our R&D is better positioned (from the 6th to the 8th place globally depending on the methods used)"* [BEY 13, p. 5]. Thus, at the European level, France is identified in the European Innovation Scoreboard 2016 as belonging to the group of "strong innovators", with innovation performances close to that of the European average, lagging behind the group of leader countries (Sweden, Denmark, Finland, Germany and the Netherlands).

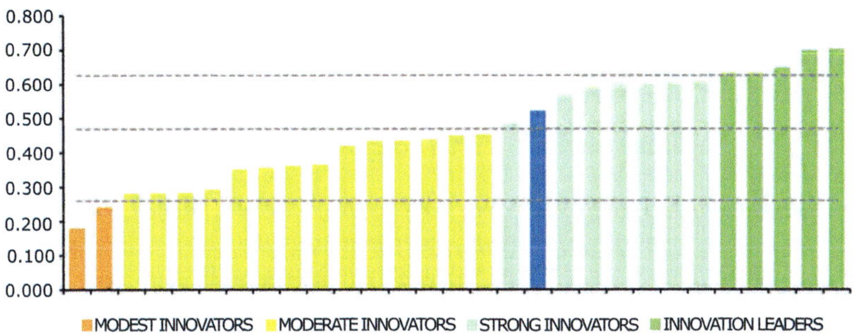

Figure 2.5. *EU member states' innovation performance (source: European Innovation Scoreboard 2016)*

This observation led J.L. Beylat and P. Tambourin to affirm in their report *Innovation: a major challenge for France Dynamizing the growth of innovative companies* [BEY 13] that this gap between the efforts in favor of R&D and performance with regard to innovation reveals the problem to be solved. The authors of the said report advocate a radical change in our way of thinking of innovation and put the role attributed to research in the right perspective: *"though innovation often requires excellent R&D, it is not limited to R&D alone. Neither is it a natural extension thereof. Innovation is above all the process which leads to the marketing of products or services* meeting a need" [BEY 13, p. 6].

While doing so, they echo the conclusions of other research works which point out that the main criticism leveled against the linear and hierarchical model of innovation pertains to the importance and the place given to R&D[10]. Innovation is presented as being completely under control, by virtue of the simplistic idea that it is enough to invest more in research in order to get more innovations. Several studies have however revealed that there is no simple causal relationship between the number of researchers, or the amounts invested in R&D, and the rate of innovation, economic growth or, more simply put, the competitiveness of a company. "*There is no statistically significant relationship between financial performance and innovation spending, in terms of either total R&D dollars or R&D as a percentage of revenues. Many companies – notably, Apple – consistently underspend their peers on R&D investments while outperforming them on a broad range of measures of corporate success, such as revenue growth, profit growth, margins, and total shareholder return*" [BOO 11]. This observation is all the more true since certain types of innovation such as marketing or social innovations do not require research. Similarly, we observe that research investment also varies depending on the industrial sector concerned.

---

10 To this first criticism, we may add: i) the linearity of the linear and hierarchical model of innovation, ii) the fact that it restricts innovation to what can be termed as technological innovation and iii) finally, the fact that this model ultimately presents each stage of the innovation process as a black box.

Moreover, data from the *Innobarometer 2007* survey show that more than 50% of innovative firms innovate without performing R&D. This statement has been confirmed by the European Community Innovation Survey [CIS 12], which underlines that almost half of innovative European firms did not perform in-house R&D, and the OECD report [OEC 10] which reveals that in Australia and Norway, the propensity to introduce a new product is similar whether or not the firm performs R&D.

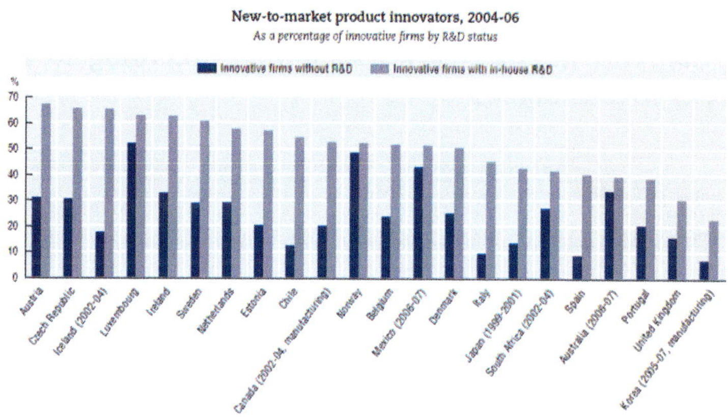

Figure 2.6. *Innovation beyond R&D (source: OECD 2010)*

This remark is confirmed in the third edition of the Oslo manual which specifies: "*while R&D plays a vital role in the innovation process, much innovation activity is R&D-based*" [OEC 05, p. 29], putting the established relationship between patents and innovation into perspective.

If the place attributed to research in innovation has to be put into the right perspective, it should be noted that the same applies to patents as well. In fact, the linear and hierarchical model provided economists in particular with a practical definition of innovation: innovation is an invention transformed into a new product or service distributed in society[11]. As such, patents have been considered as an indicator of innovation. However, a study conducted by R. Fontana, A. Nuvolari, H. Shimizu and A. Vezzulli [FON 13] on a corpus of 3000 innovations rewarded over the period 1977–2004 clearly challenges the established relationship between patents and innovation. In fact, their study reveals that only a small number of these innovations were based on a patent. To be precise, they note that only 10% of the rewarded innovations were based on a patent and specify that this number varies somewhat according to the industry type considered.

	Awarded innovations	Patented innovations	Share not patented
All the sample (1977-2004)	2802	255	90.9%
Non corporate	886	25	97.16%
Corporate only	1751	220	87.44%

Figure 2.7. *Number of patented innovations reported in the total number of rewarded innovations over the period 1977–2004 < 10% (source: [FON13])*

This analysis was corroborated by P. Fauconnier who underscores that if we need proof of the gap between the innovation model and the actual ability to innovate:

> *"it would suffice to see that both Alcatel Lucent, which is preparing to sack 10,000 employees and shut two sites in France, and Blackberry, which has just had a brush with bankruptcy and sold itself in a hurry to an investment fund, have been ranked pretty high in the Thomson Reuters ranking list. Similarly we can see that the French*

11 According to the third edition of the Oslo manual, diffusion is the way in which innovations spread, through market or non-market channels, from their very first implementation to different consumers, countries, regions, sectors, markets and firms. Without diffusion, an innovation has no economic impact [OEC 05, p. 17].

> *company which filed the maximum number of patents in 2012 is PSA Peugeot Citroën, with 1,348 submissions, whereas this very company found itself in a more than critical situation this year registering losses to the tune of 5 billion euros*". [FAU 13]

It thus seems that France is better at making inventions with money than making money with inventions.

This analysis is even more problematic because as R. Boschma [BOS 08] points out that there is still a strong belief in the EU that R&D policy will bring benefits to many regions. However, most of the new knowledge created is not exploited economically in Europe but is being used in other countries like the United States. Thus, from 2004 onwards, 90% of French transactions with regard to patents have involved sales as against only 10% of acquisitions, an imbalance pointed out by the Cour des Comptes (CDC – Court of Audits) in its report on public funding of research published in [FAU 13]. This means that the European R&D policy nurtures the exploitation of knowledge elsewhere [BOS 08].

The preceding conclusions lead a number of observers to emphasize, on the one hand, the gap between the theoretical notion of innovation and the reality of companies [MOR 09] and, on the other hand, that in spite of an explosion of the number of national programs in favor of innovation[12] in France, it continues to be difficult to speak of a veritable research and innovation policy. In fact, as noted by P. Laredo [LAR 11], in the course of the first decade of the 21st Century, European policies have culminated in a true Europe of higher education and research. On the contrary,

12 The State and its operators were managing close to 30 national programs in 2000. Their number then rose to 62, to which the programs managed by local authorities must be added [PIS 16, p. 8].

there has been hardly any progress made with regard to the ambitions of Lisbon centered on a Europe of innovation. Let there be no misunderstanding on what we are saying, we are not trying to undermine the role of research in the dynamics of innovation in any manner whatsoever but only to relativize the place attributed to it in the linear and hierarchical model. In fact, although R&D is an essential activity, it is however not enough. Even though innovation originates from scientific advancements, as we will see in the following sections, there is a whole series of stages, including design, which will have to necessarily follow it. This is not a new idea. Among the first authors to have explained the design process in the innovation process are S. Kline and N. Rosenberg[13] and we now propose to present their model.

2.8. The design process at the core of the innovation process

The particularity and richness of the Kline and Rosenberg model (Figure 2.8) is due to two things:

– it highlights the idea that the innovation process should not be understood and represented from a single axis (from the scientific concept to the new product made available on the market), as the linear and hierarchical model of innovation does, but must be represented from five specific paths, which will be examined in detail subsequently;

– it shows that design plays a central role in the innovation process, which makes it possible to create a link between laboratory activity, the activity of industrial services (industrialization, methods, etc.) and production.

13 S. Kline (1922–1997) was an emeritus professor of mechanical engineering and N. Rosenberg (1927–2015) was an American economist specializing in the history of technology. The chain-linked model or Kline model of innovation was introduced by S. Kline in 1985, and further described by S. Kline and N. Rosenberg in 1986.

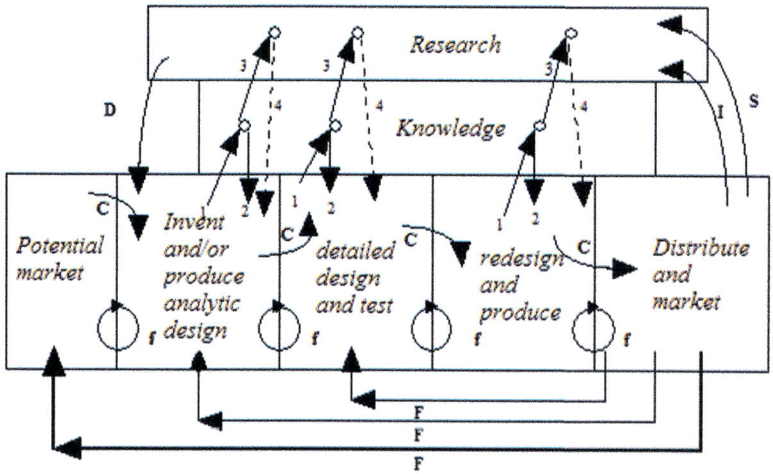

Figure 2.8. *The chain-linked model (adapted from [KLI 86])*

As we said earlier, the innovation process cannot be represented from a single axis according to the authors. Five paths can be identified:

1) The first path, the central axis, places the design process at the core of the model, and therefore at the core of the innovation process. With regard to the linear model of innovation, this is a fundamental point of departure, insofar as the authors reject the widespread view of the omnipotent nature of science in the act of innovation by making the design process the backbone of their model: "*the central process of innovation is not science but design*" [KLI 86, p. 286].

For S. Kline and N. Rosenberg, the design process consists of a succession of stages that represent, respectively, invention and analytical design, detailed design and testing, and the final design.

2) On this first axis, a second path juxtaposes, which represents all the feedback loops that can:

– come about between two successive stages of the central chain, which is identified by the letter C;

– go back several stages earlier;

– act retroactively on the innovation process as a whole, by creating a new need for instance.

Such feedback loops are the mainstay of any innovation process. They are indicated according to their effects by the letters f and F.

It may be noted that these first two axes specify two things:

– the design process is necessary to initiate an innovation; in other words, there is no innovation without design;

– re-design is essential to get an actual innovation after a number of iterations.

3) We have seen above that the innovation process approach initiated by science is not robust. Science is sometimes solicited during a process for a problem that had not been envisaged at the outset: "*over the course of history thus far, it is moot whether science has depended more on technological processes and products than innovation has depended on science... In his work on the electric lighting system, Edison was forced by the needs of the system to pay a mathematician to work out the analysis of the parallel circuit*" [KLI 86, p. 287]. It is on the basis of this observation that S. Kline and N. Rosenberg have chosen to include scientific intervention in parallel with the central chain in their model in order to emphasize the fact that it is

present throughout the process. The author seeks to however clarify two points:

– the use of science can take two forms, depending on whether the process initiated by the central axis requires drawing from a stock of available knowledge (K-link), or solicits updating or creation of new knowledge (R-link) when this stock does not suffice to solve the problem posed;

– the nature of the scientific knowledge sought varies according to the stage of the central chain in which the designer finds himself. If the results of fundamental research are used during the invention phase, it appears that, during the development phase, there is a greater interest given to research related to the manner in which the components of a system interact (experimental research based on trials and tests). Finally, in the production stage, it is most often research devoted to the production process that is solicited.

These multiple relations with science are expressed by the K and R links.

4) As far as the D-link is concerned, it illustrates radical innovations, albeit much more rare, which originate directly from the development of new sciences. This is the case of semi-conductors.

5) Finally, the last path of the model, represented by the I and S links, represents the feedbacks that can come from the innovation as an outcome of scientific dynamics. "*Without the microscope, one does not have the work of Pasteur and without that work there is no modern medicine*" [KLI 86, p. 293].

What we must retain from the preceding is that Kline and Rosenberg's model was the first model which underscored the role of design in the overall innovation process. It emphasizes that the improvement of the innovation process will depend not only on the number of researchers employed and the budget allocated to R&D, but also on the interactions between the different stakeholders, the effectiveness of cooperation and the way in which the design process is carried out. The issue that comes up now is to understand what precisely we mean by design.

2.9. The design process, what are we speaking about exactly?

The models of the design process are multiple in scope. N.F. Evbuomwan, S. Sivaloganathan and J. Jebb counted around 20 in an article titled *A survey of design philosophies, models, methods and systems* [EVB 96] and L. Blessing found even more of them in his thesis titled *A Process-Based Approach to Computer-Supported Engineering Design* [BLE 94].

However, according to L. Blessing it is possible to distinguish two approaches in the literature, and hence two models of the design process, namely the Anglo-Saxon model and the German model.

2.9.1. *The two models of the design process according to L. Blessing*

The first model, the Anglo-Saxon model, is guided by the management of the design process, hence the extreme attention paid to the needs of the consumer and the cooperation of the stakeholders of the process. It is also a product-centered approach, as the concept generation process is based on the analysis of the initial idea of the product. As an example, L. Blessing cites the L. Archer [ARC 65] model.

L. Blessing contrasts this model with the German model of G. Pahl and W. Beitz [PAH 84]. The latter corresponds more to a methodical approach in order to improve the efficiency of the design process itself. It is a more abstract approach in the sense that it is primarily problem-oriented. In fact, in the second model, each stage of design is characterized by a different relationship to the object. The object is sometimes represented virtually, functionally, sometimes by an architecture of components, etc. From this it appears that the more we advance in the process, the more concrete and quantitative the results are, and the more the variety of alternatives decreases. L. Blessing emphasizes, however, that since the model of G. Pahl and W. Beitz is problem-centered, emphasis is not given to the analysis of needs which comes within the scope of a product-centered approach. Similarly, the constraints related to the product design environment are not taken into account.

In reality, these two models seem to be more complementary than conflicting. It appears to us that the description of the process that we will be proceeding with in the next section contributes greatly to their reconciliation. However, we agree with L. Blessing when she asserts that this different representation of the design process seems to be able to explain why the ways of improvement of the design process in these two groups of countries are different.

2.9.2. *The stages of the design process*

Most design theoreticians and practitioners agree to define the design process as a succession of four stages:

– The "clarification of task" or "design specification" stage aims first at understanding and explaining the need. In other words, it consists in defining the problem. Second, it leads to defining the design environment. The designer must then distinguish the objectives (the required functions, the final cost of the product, a quantity, etc.), the permissions (whether it is a fixed demand, a minimum demand, or a wish; in other words, what is the extent of freedom that the group of designers have?) and the constraints of the artifact (it can be an issue relating to resistance, corrosion, manufacturability, compatibility, i.e. insertion of the artifact in a socio-technical environment, etc.). It is during this first stage that the vector of objectives and constraints is initiated.

– The "conceptual design" stage implements a functional analysis approach, this time applied to the artifact to be designed. It proposes to:

- list the different functions of the artifact to be designed;

- look for possible solutions for each of the defined functions;

- proceed with the assessment of possible combinations.

– The "embodiment design" stage is aimed at materializing and dimensioning the artifact being designed. During this stage, we generally distinguish:

- the "layout design" or design of the arrangement of the functions and parts into a consistent and compatible set;

- the "form design" or design of the external appearance of the arrangement.

– The "detail design" step generates the manufacturing instructions, owing in particular to the drafting of specifications, which include all the specifications for the production (techniques retained, shape and dimensions of the various components, etc.).

2.9.3. *Overall convergence of the design process*

Each stage of the design process has as input a vector of objectives and constraints which is specific to it and produces a new vector of objectives and constraints at the end of the stage, which then constitutes the entry vector of the objectives and constraints of the next stage.

This is why we will talk about the filiation and propagation of a vector of objectives and constraints in the sequence of the various stages of the design activity. Filiation because the objectives of the subsequent stages specify the objectives of the earlier stages. Propagation because the objectives of the subsequent stages cannot contradict the objectives of the earlier stages.

As the degree of freedom in the expression of the vector of objectives and constraints decreases overall as the design project moves forward, the space of the problem, linked to the definition of the artifact and not to its production, tends to decrease, hence we get a design process linking the stages more and more quickly.

It is therefore possible to represent the design process by a model said to be "conical" (Figure 2.9).

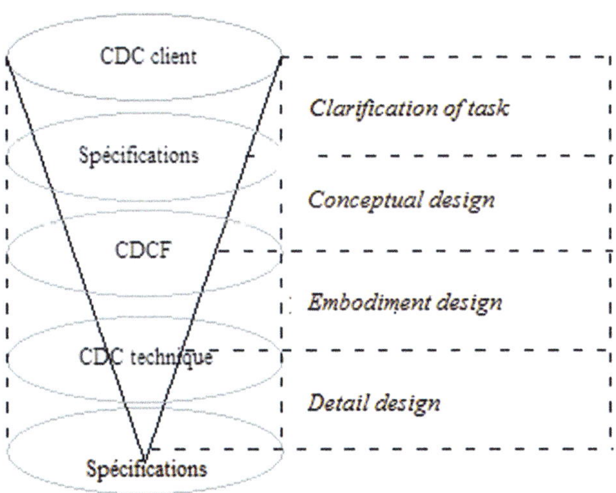

Figure 2.9. *Conical model of the design process*

What makes such a representation convenient is the fact that it makes it possible to emphasize that the design process is a multi-stage process and its nature is convergent on the whole. Its disadvantage lies in the fact that it seems to suggest a certain amount of linearity in the design process. However, the study of design projects shows that this is not the case and that there are many feedbacks during the process due, on the one hand, to the design review procedures and, on the other hand, to the adoption of opportunistic problem-solving strategies by designers owing to the cognitive cost of solving the problem they are faced with.

Finally, it should be pointed out that the "thickness" of the design process may vary depending on the type of innovation project. To be more precise, the more innovative the project, the greater the "thickness" of the design process. As an illustration, in the case of an improvement project, designers devote far less time to the stage of the process that we have named clarification of task, as well as to the conceptual design stage when the functions and principles of solutions are already known. Innovative projects will devote proportionately more time and energy to the first phase of the process, whereas integration projects will spend proportionately more time and energy in stages 2 and 3 of the design process.

2.9.4. *Rule-based design regime versus innovative design regime*

If design is the central process of the innovation process, it should be emphasized that companies do not organize design in the same way depending on whether they work in a rule-based[14] design regime or in an innovative design regime, to use the terminology developed by A. Hatchuel, P. Le Masson and B. Weil [HAT 06].

Rule-based design stems from the emergence and development of design offices in the late 19th Century. Since management does not like anything vague and uncertain, this form of organization has made it possible to streamline design. The transition from the specifications stage to the proposal of a solution is organized according to a programming logic. In the rule-based design regime, a stable corpus of rules builds reasoning, organization and performance. The organization of rule-based design regime combines a strong expansion (variety of products designed) by reusing the existing knowledge to a maximum extent. Such a form of organization of design has shown its effectiveness in innovating based on parametric performance criteria (faster, cheaper and safer) and therefore in enhancing an object within the framework of a given identity. It would, however, prove to be unsuitable for the emergence of innovations in which the very identity of the object is revisited. This is the reason for the development of the innovative design regime.

In fact, innovative design causes recurrent identity crises of the artifacts (the iPhone revisits the identity of the phone, and it is the same case with the Dyson bagless vacuum cleaner or the Actifry fryer from Seb which revisits the identity of the fryer) and confuses borderlines (where is the borderline between a food item and a medicine today?). However, redefining the identity of artifacts implies that designers

14 By design regime, the authors refer to a form of industrial design, the latter being a form of collective action characterized by:

- design reasoning;
- a form of collective organization;
- a rationale of performance.

must carry out innovation processes whose aim is to generate expansive partitions, producing a dual expansion in concept and knowledge.

2.10. Validity of the model

The importance of the role played by the design process in the innovation process has been confirmed on multiple occasions. And so it was in the *Made in America* report [DER 10] drawn up by an interdisciplinary commission created in 1986 by MIT. The commission, comprising 16 academics having hands-on industry experience and including renowned economists like R. Solow, reached the observation that one of the main ills afflicting the United States in the 1980s could be attributed to the fact that they had difficulties in transforming their inventions into new products. The *Made in America* report concluded as follows: "*investing in fundamental scientific and technical research is absolutely essential for sustainable economic growth... however scientific feats do not automatically end up with commercial success. For achieving this, new ideas need to have been converted into products: the products that the customers want, when they want them and before competitors become capable of providing customers with such products. And these products must be manufactured efficiently and they must be of high quality*" [DER 90, p. 90].

The importance of design in the innovation process was also established in the late 1980s by the former IBM Vice-President of Research. Based on a comparison of Japanese and American market positions, R. Gomory emphasized that leadership position in a product can be acquired without scientific leadership, as long as firms excel in design activities [GOM 89]. This observation led him to conclude that American companies should give priority to change the way they think design since, as he specified, when competition between companies is based on innovation, it is not won by doing more science, but by improving the product development process.

A survey of Swedish companies showed that innovative companies are more likely than non-innovative companies to regard design as a strategy.

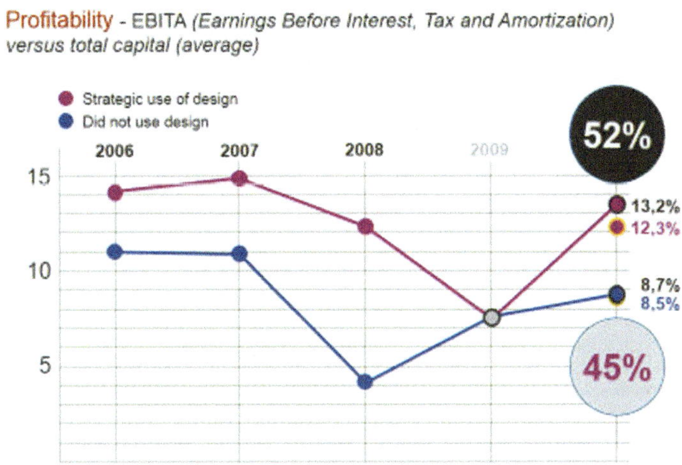

Figure 2.10. *Companies that invest in design are more profitable (source: Swedish Industrial Design Foundation). For a color version of the figure, see www.iste.co.uk/forest/innovation.zip*

A 2012 study showed that, on average, for every £1 businesses invested in design, they gain over £4 net operating profit, over £20 net turnover and over £5 net exports. More broadly, businesses that invest in design have approximately 50% better long-term financial performance than businesses that do not [DES 15].

In 2009, the European Commission report design as a driver of user-centered innovation recalled that "*firms that use their design activity as a strategic driver are five times as likely to develop new products as compared to firms that do not work consciously with design*" and that "*a survey of Irish companies showed that 75% of SMEs that use design engage in what the survey categorises as the most radical type of innovation — developing new products and services for new customers. This compares with 48% of companies that do not use design*" [COM 09, p. 15].

The importance attributed to design is finally perceptible among manufacturers who now see design as a key element of the innovation process and development of new products (Figure 2.11).

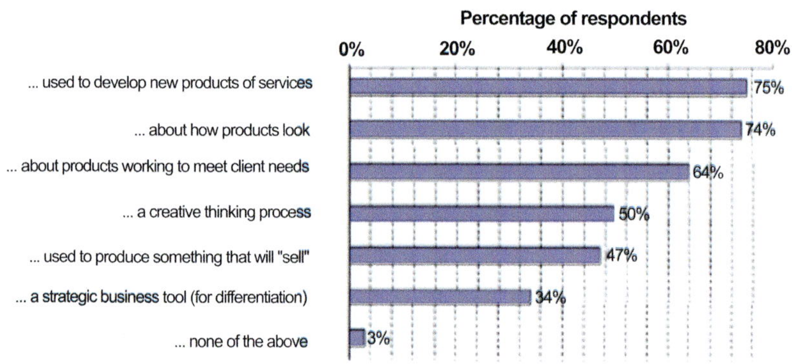

Figure 2.11. *What is design? (source: Tether (2005) in [COM 09, p. 14])*

This awareness translated into a significant change in the positioning of Danish companies on the design ladder between 2003 and 2007[15]. Although 36% of companies did not seek recourse to design in 2003, they represented only 15% in 2007. Conversely, design has become strategic and a driving force of innovation for 21% of them as compared to 15% in 2003 (Figure 2.12).

15 The Design Ladder was developed by the Danish Design Center (DDC) in 2001 to measure the extent to which companies use design.

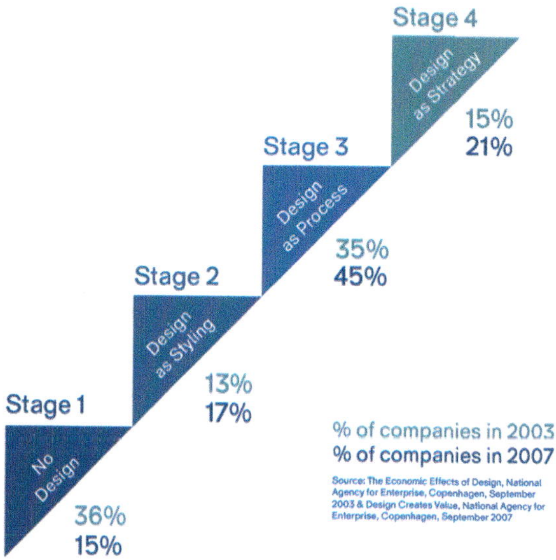

Figure 2.12. *Evolution of the place of design in companies (source: [MEL 10]). For a color version of the figure, see www.iste.co.uk/forest/innovation.zip*

The preceding pages have allowed us to put the place given to research in the right perspective and reveal the central role of the design process in the innovative phenomenon[16]. By doing so, they invite us to rise above the linear and hierarchical model of innovation and to think of innovation differently.

In fact, the model of S. Kline and N. Rosenberg makes it possible to understand that the relations between science and innovation are much more complex than what the Schumpeter model presumes them to be. In some cases, basic research work initiated independently from a pre-existing need what led to innovations. On the contrary, sometimes an

16 It should be noted that, following the example of Finland, Denmark and the United Kingdom, a few countries have taken a head start by focusing their policies on design for several years now.

innovation is the one that leads to new research. As pointed out by H. Brooks [BRO 94], "*one of the most dramatic examples of the generation of a stimulus to a new field of basic research by a discovery made in the course of a technology-motivated investigation was the discovery and quantitative measurement by a Bell Laboratories group in 1965 of the background microwave radiation in space left over from the original 'big bang', for which Penzias* et al. *ultimately received the Nobel prize*" [BRO 94, p. 482]. Thus, to establish a causal relationship between research and innovation, as the linear and hierarchical model does, is an erroneous representation because research opens a whole host of possibilities to innovation just as the latter opens all the possibilities that research holds. In his inaugural conference of the Technological Innovation Chair at the Collège de France, D. Roux perfectly illustrated such a point of view[17].

If we accept the idea according to which design plays a central role in innovation, we must fine-tune S. Kline and N. Rosenberg's model and increase our knowledge on the design process. In short, we must adopt what we have named, with J.-P. Micaelli, the artificialist perspective [FOR 03].

17 D. Roux's inaugural lesson is accessible at the following link: https://www.college-de-france.fr/site/didier-roux/inaugural-lecture-2017-03-02-18h00.htm

3

Artificialism

Artificialism is a theory that puts design at the core of human action [FOR 03].

Artificialism is a theory that pertains to artifacts, by which we mean all the objects that are designed by man in order to meet his needs. Although Artificialism cannot explain why, at a given time and place, a particular value, a particular conflict, a particular social protest movement or a particular "macroactor" (a professional, political or social group, etc.) appears, it can explain, on the basis of the needs arising from these values, how the concerned groups design and produce alternative artifacts.

Artificialism is in fact centered on the tactical dimension of action, i.e. on the process that ensures the transition from the formulation of a need to the realization in finite time of satisfactory solutions. We are thus led to the observation that the reversal of perspective that Artificialism calls for, although seemingly insignificant, is important since it pertains to thinking of the world of the artificial as an outcome of the way in which man approaches his external environment in order to act on its components in such a manner that they fulfill his needs [SIM 69]. In this way, the artificialist "*knows with certainty the why and the how of its realization, instead of being confined to hypothetical speculations on the origin of natural objects: transformism, evolutionism, genetics, natural selection, divine creationism*" [LEM 99, p. 157].

Nevertheless, before presenting the five key propositions that establish Artificialism as a theoretical framework and their implications, we feel that it is important to briefly present the Simonian conception of artificiality, which makes up the basis of Artificialism.

3.1. Artificial world as a set of artifacts

Simon starts from the observation that man lives in an anthropized world made up of artifacts. By artifact, he means a generic category, which comprises a number of entities:

– appearing in the form of objects, codified physical techniques, environments such as gaming devices, virtual reality, organizations and urban environments;

– unique and singular or multiple and commonplace (a common consumer good like a drilling machine);

– short-lived or lasting;

– tangible (the components of the drilling machine) or not (the software of its speed regulator);

– abstract (its functional model) or concrete (its development);

– one-piece (its half shells) or structured (its inner modules);

– software intensive (the virtual reality environment shown in Table 1) or not (the chuck of the drill of the drilling machine);

– nanoscopic (therapeutic nanoparticles) or macroscopic such as technical languages, macro-networks such as the Internet, etc.

– occasional (the drilling machine) or resilient (the power grid to which it must be connected to be supplied with electricity);

– passive (its shell) or active (its speed regulator);

– positively connoted (medical treatment) or not (anti-personnel land mine), etc.

H. Simon takes up the world of the artificial by adopting a global theoretical approach. He does not content himself with merely identifying the artifacts created and used by man, as the encyclopedists did for instance, and thus defined all the artifacts as an extension (i.e. a collection). He defines it as an intension, meaning thereby any entity designed by humans to satisfy their needs irrespective of the concrete form it takes.

To put it precisely, if the world of the artificial exists, it is because man cannot satisfy all of his needs by having a passive attitude or that of an outright predator or consumer with respect to nature [FOR 03, p. 24]. Thus, if an artifact exists, it is because it is designed to meet needs, and H. Simon specifies that artifacts adapt themselves to man's goals and intentions: when they change, his artifacts change as well [SIM 95].

Therefore, to understand what an artifact is, it is first and foremost necessary to understand what it is made for. Its primary objective attribute is its function. Its criterion of adaptation is the level of adjustment to a need. In pursuance thereof, the performance and dynamics of the artifact must be understood, from an objective point of view, on the basis of its functional (does it always meet the need?) or dysfunctional (what dissatisfactions do its existence or its usage generate?) capabilities. The practical corollary of this assertion is that a "good" designer is not the one who seeks to express his skill in producing perfect things, but functional ones with regard to particular needs.

The second pillar of Simonian conceptualization is the design process. However, in order to be qualified as an artifact, the latter has to not only respond to a need but it is also necessary that the response to this need has necessitated a synthesis or what is commonly called a design process.

In fact, according to H. Simon, in order to be qualified as an artifact, the considered entity must respond to a need and satisfy a function. It is also necessary that the response brought with regard to this need has necessitated a synthesis developed by an ideal-typical actor called designer. This point is often overlooked by anthropologists or philosophers of techniques. According to them, for example, blocking with a pebble the wheel of a car whose parking brake is no longer working and which is therefore likely to run down a slope, makes that stone an artifact. From a Simonian point of view, it is not an artifact[1]. However, if we ask how to prevent or compensate for such a malfunction and create several objects in response to this function, it is then that we create an artifact called a wedge. Thus, design supposes a conceptual detour: it should not be confused with "usage".

1 On this point he agrees with the philosopher of technology G. Simondon: an artifact is not only something that we use, because some use can be found for anything and everything.

3.2. Contribution of the Simonian theory to the understanding of the design process

Understanding H. Simon's point of view involves getting back to his theory of decision for a moment.

3.2.1. *Bounded rationality and satisficing*

Since 1934, H. Simon had always been very interested in the rationality of decision behavior within public administrations. He observed that decisions are not made the way standard theory suggests. H. Simon aimed at understanding how decisions are really made. To this end, he borrowed a methodology from behavioral economics which requires the analysis of the behavior of individual actors. He thus radically broke from the marginalist analysis and presented the rationality of action from the decision-making process leading to action. To be more accurate, H. Simon refuted the hypotheses of pure and perfect information and of perfect rationality. He therefore also rejected the idea of the omniscient decision maker (homo œconomicus) and promoted the concept of bounded rationality.

The concept of bounded rationality, he formally defined in 1957, explains the following limits of the decision maker faced with a problem:

– an imperfect or limited knowledge of the decision maker's environment;

– the impossibility of anticipating and considering all options to solve a problem, due to limited abilities of calculation;

– the impossibility of processing all available data, due to the limitations of attention.

According to H. Simon, considering the bounded rationality of the decision is important but is not enough. How the decision is really made should also be understood. Simon claimed that decision-making starts when a need for decision is identified and that the next step consists of seeking alternatives as an answer to this need. One of the main components of the research procedure of alternatives is the "stop rule". Indeed, there is seldom the guarantee that, on the one hand, any optimal solution can be found among the alternatives and that, on the other hand, an optimal solution can be found as long as the research goes on. A standard should also be given regarding a satisfying solution, thus defining an aspiration level.

How does this aspiration come to be defined? Based on his past experience, the decision maker forms a judgment of the quality of the solution he can aspire to achieve, with an investment deemed reasonable. This defines an aspiration level and thus determines when the research should stop. Taking into account the adjustable nature of the aspiration level helps to explain why a decision considered from the viewpoint of the research procedure can be described as satisfactory [FOR 01].

3.2.2. *Design as a process obeying satisficing*

As early as 1959, H. Simon presented the design process as a constrained problem-solving process[2] that would obey the principle of satisficing: "*designing is satisficing, finding an acceptable solution*" [SIM 95, p. 246]. In fact, he specifies that the purpose of the design process is not to

2 Note that defining the design process as a problem-solving process avoids reducing it to a decision; the decision is only a moment in the process just as choice is only a particular moment in the decision.

maximize any constraint, but rather to satisfy a certain number of parameters:

> *"In solving problems by heuristic search, a criterion is needed for deciding when a satisfactory alternative has been found. (...) For example, the criterion for a new automobile design may require that it be manufacturable at less than a specified cost, that its gasoline consumption be less than a specified amount per mile, that its exhaust gases meet the legal standard of emission, that it incorporate safety belts, that it have a 'streamlined' appearance and chrome hubcaps, and so on. The list can be as long as we please, and there is no need to specify how much of one of these requirements would be traded off for a given improvement in another. All have to be satisfied in a satisfactory design. (...) If alternatives cannot be found that satisfice, then aspiration levels will drop until an alternative is found"* [SIM 92, p. 30].

Three empirical reasons confirm the satisfactory and non-optimizing nature of the design process. These are:

– Heterogeneity and contradictions between objectives. As the great Finnish architect A. Aalto (1898–1976) indicated:

> *"Let us assume that we have to build a church. The quality of the soil, the geographical location and the topography, the material used (...) the heating system, the ventilation, the lighting, the treatment of the external surfaces and countless other factors (...) even if they are all parts of the same building, we often see that there are contradictions between them, and they must be*

> *reconciled in order to achieve harmony (...). In all the tasks (...) there are tens or even hundreds and sometimes thousands of contradictory elements that can be harmonized only by the human will"* [AAL 88, p. 169].

– The lack of certainty as regards the optimality of the solution. Designers are never sure of finding the best solution, owing to:

- the complexity of the artifact;

- the variety of possible technical solutions;

- the available knowledge;

- and, above all, the strong time constraints which cannot really be negotiated and weigh upon the duration of the design project.

– The importance of compromise. In fact, every artifact should be seen as the result of a compromise between:

- the constraints related to the definition of the problem, inherited by the user, the consumer and the designer;

- the constraints related to the design environment (standards, limitations of knowledge, etc.) that are imposed on the designer;

- the inherited and propagated constraints of the upstream phases, expressed in the vector of objectives and constraints, derived from the problem resolution status;

- the constraints that emanate from the actors of the design process, whether the reference made here pertains to the issue of the limited rationality of the economic actor raised by H. Simon or to their assessment of the situation. As pointed out by P. Falzon, relying on a study of the design processes of traffic junctions, experts confronted with the same problem can produce different results:

> *"The same design problem was submitted to six domain experts. The designers were asked to verbalize their reasoning aloud and the traces of the work were collected (graphs, notes). We can see that the six individuals produce six different solutions. It must be understood that this observation conveys neither an experimental artifact, nor unequal competences nor the lack of a rigorous reasoning process. This is a characteristic of design situations, linked in particular with the weighting of the constraints, which varies from an individual to another: this weighting leads different subjects to elaborate different representations of the problem itself. The individuals are not 'playing' the same problem"* [FAL 90].

From the above, it appears that:

– by definition, any artifact should be considered as the best solution found at time "t", taking into account the past (especially the knowledge available and the existing artifacts), the place (here we are referring to the role of institutions and routines) and the individuals who participated in its design;

– the artifact is coherent not because of a natural convergence and even less because of a miraculous combination, but because of the designer's activity.

3.2.3. *Specificities of the design process*

If, according to H. Simon, the designer engages in a process that inherits the properties of the decision and the problem resolution, he is equally aware of the specificities of the design process.

He notes that the *"design problem"* if often an *"ill-defined problem"*, which implies that all of the data is not known right away and therefore the rules are likely to change.

H. Simon specifies furthermore that design is a process of problem forming, problem finding and problem solving[3]. In fact, analyses of designers' actions show that designers not only have to define the desired functions (referred to as "problem forming" by [SIM 95]), they also have to spend considerable time and energy generating acceptable alternatives, i.e. seeking an internal architecture that satisfies the functions for which it was designed. In other words, by designing, we learn what we want[4]. H. Simon illustrates this with the parable of oil painting. While making an oil painting, each new touch of color put on the canvas creates a kind of organization that provides a source of new ideas to the painter. The act of painting is a process of cyclic interaction between the painter and the canvas, in which current objectives lead to new applications of painting, while the gradually changing organization of the painting suggests new objectives. Therefore, design is a process through which, on the one hand, the objectives (as the design problem is continually reformulated during the design process) emerge and, on the other hand, the alternatives (as the objectives get stabilized only towards the end of the process, according to H. Simon) emerge.

3 The design process cannot therefore be reduced to choice.
4 Design problems are problems said to be "ill defined" [SIM 84].

3.3. Simonian empiricism

Considering artificial objects from an artificialistic perspective leads to adopting a posture that is not neutral:

1) It is a matter of asserting that knowing an artifact or an innovation means knowing its process of design, which, in other words, leads to emphasizing that the object of knowledge is not the "object" (the artifact in its existential reality) as such but rather the "process":

> "*The artificial world is centered precisely on this interface between the inner and outer environments; it is concerned with attaining goals by adapting the former to the latter. The proper study of those who are concerned with the artificial is the way in which that adaptation of means to environments is brought about – and central to that is the process of design itself*" [SIM 69, p. 68].

In the wake of the Simonian approach to the artifact, the artificialistic perspective compels us to adopt a significant decentralization, which consists of questioning the artifact from the point of view of its generating process, namely the design process. The change of focus that Artificialism calls upon impels us to study the process (design) more than the object (its result), the interactions and transactions between designers and successive users more than its lineage, etc.

The study of artificial objects, and more precisely of the process of the design of said objects, necessitates seeking recourse not to the deductive method – often assumed to be the only scientific method – but seeking recourse to an alternative method, namely empiricism[5]. In fact, an artifact can be defined empirically. All that is needed is to associate a designer[6] and a design process or a redesign (rationalization) process with an object in order to designate it as an artifact.

2) It is also a matter of asserting that there is no metaphysics of the artifact, i.e. something that would explain its "raison d'être" outside of human understanding and human action. As a result, *"it is no longer legitimate to restrict the field of scientific disciplines to the knowledge of natural phenomena alone: in principle, artificial sciences (Simon, 1969; 1991) produce as much scientific knowledge as 'natural sciences' that educational institutions have been accustomed to for two centuries"* [LEM 99, p. 93].

3.4. Key propositions of Artificialism

In 2003, in *Artificialism: Introduction to a Theory of Design* [FOR 03], we proposed a first attempt at theorization of Artificialism, an attempt which systematizes the remarks of H. Simon on the basis of the following five proposals:

PROPOSITION 1.– Artifacts are universal, meaning that they are necessary as soon as any perceived need cannot be satisfied immediately, by collecting it from nature, buying it from a market or by applying customary habits. Such a definition does not, therefore, reduce the artifact, nor

5 In spite of the numerous criticisms addressed to him, H. Simon never denied his logical positivism, simply specifying in his book *Models of my life* that he would now call it logical empiricism.

6 It should be noted that the notion of designer does not mean that he is unique (most of the artifacts are the fruit of collective work) but rather refers to an "ideal model" made necessary by Artificialism. It should also be noted that Artificialism rejects the image of an omniscient and omnipotent designer [FOR 03, p. 41].

innovation which is of more particular interest to us in this book, to a technical object alone.

PROPOSITION 2.– The existence and fate of artifacts result from design[7]. This is carried out by an individual characterized by the ideal model of the designer[8] who takes care of all or part of it. In doing so, Artificialism makes it possible to emancipate oneself from "the tendency of many individuals to consider technology solely in terms of enslavement".

As the philosopher of technology G. Simondon pointed out as early as 1958: "*The opposition made between culture and technology, between man and machine, is false and baseless; it only conveys ignorance or resentment*" [SIM 89, p. 9]. According to him, because culture blocks out technical objects in a world that has no meanings but only a use, it is ignorant of the human reality contained in all artifacts:

> "*The behavior of culture towards the technical object is akin to the behavior of man towards a foreigner when he allows himself to be carried away by primitive xenophobia. Misoneism directed against machines is not so much hatred of a novelty as a rejection of the foreign reality. However, this foreign being is still human and the comprehensive nature of culture is what makes it possible to discover the foreigner as being human. In the same way, the machine is the foreigner; it is the foreigner in which is enclosed the unknown, materialized, enslaved humaneness, but which*

7 As a good pragmatist, the artificialist in fact knows that designing is never final, and that the very idea of an ideal and final artifact is meaningless. A final artifact is not only a response brought about to a given need at a given point in time, but it is also a set of problems left for future designers to deal with [SIM 95].

8 We are talking about the figure of the designer but are aware of the fact that, in reality, behind this ideal model hides a collective of stakeholders.

> *still remains human. The greatest cause of alienation in the contemporary world lies in this ignorance related to machines, which is not an alienation caused by the machine, but by the lack of knowledge of its nature and its essence, by its absence from the world of meanings, and by its omission in the table of values and concepts which makes up an integral part of culture.*" [SIM 89, p. 9–10].[9]

On the contrary, by putting emphasis on the ideal model of the designer right from the outset, Artificialism not only makes it possible to reveal the human dimension of artifacts but also to consider that understanding the mode of existence of artifacts is akin to understanding our way of relating to the world and to giving us the means to think of ourselves[10].

The designer acts by initiating a process of design, or re-design, by linking both design goals and strategies:

– design goals, which may be external: the designer must ensure that the artifact meets the different requirements and constraints (e.g. interoperability) specific to the need that it must satisfy. The goals can also be internal. The designer of the artifact must not only identify the expected

---

9 This observation was used in 1983 by J. Lang, then Minister of Culture in France, when he deplored the fact that people could actually make a distinction between "culture in clean hands" and "culture in dirty hands", pointing out that there is a certain reluctance to consider that the action of inventing can be carried out at the level of a noble and major activity. We have argued that such a position is inherited from the lack of technical culture [CHO 17].

10 Technical culture makes it possible to understand that during their genesis, innovations/technical objects do not escape from validation, from diffusion in society which has witnessed their birth and from history. Note, however, that this does not mean that we restrict ourselves to a deterministic reading because design clearly explains the jumps, nor to a vision of social indeterminism because of the cultural dimension.

functions of the artifact to be designed (the need) but also i) integrate components that are both very numerous (10^4 pieces for a car or a helicopter) and heterogeneous (materials, organs, networks, etc.); and (ii) arrange this assembly in a coherent and satisfactory manner.

– design strategies that integrate both the dynamics specific to what designers do and what companies want, and which translate into a locking (synonym: closure) or an opening artifact (Figure 3.1).

<table>
<tr><th>Design goals</th><th>Design strategies</th></tr>
<tr><td>Taking into account environmental constraints in aircraft design</td><td rowspan="2">Technical opening: opening up future design opportunities. In the context of a call for tenders, specifying the functions of the artifact while giving the respondent free choice to develop the technical details of the solutions.</td></tr>
<tr><td>Development of smart grids based on two new interoperability standards between devices and terminals and smart grids launched in 2012 by the European Telecommunications Standards Institute and the Energy Services Network Association for the European Union area.</td></tr>
<tr><td>New competing artifacts: substitution of the watch for the smartphone.</td><td>Technical closure: restricting future design opportunities.</td></tr>
<tr><td>Improvement: use of magnets to enhance the mass and efficiency of the drill motor.</td><td>Opening to community design: free software, participatory urban design.</td></tr>
<tr><td>New architecture: move from central engines to distributed engines (tramway, metros, etc.).</td><td>Monopolistic locking: rigid interface between the boots and ski binding. Buying one will mean buying the other.</td></tr>
</table>

Figure 3.1. *Artificial dynamics (source: [FOR 07])*

PROPOSITION 3.– Design is a restricted process from a temporal point of view. It is:

– creative: we cannot predict its outcome or its progression;

– proactive: we cannot design without producing intermediate artifacts;

– evaluative: designing means evaluating the performance of what is being designed and how it is being designed;

– complex: different views and memories are necessary for designing.

Even if it includes both temporary and local choices, design does not amount to just a series of choices or the application of unconscious routines. The designer will always have a singular problem to deal with. As a result, understanding the mode of existence of artifacts will involve revisiting the issue of creativity.

PROPOSITION 4.– The design process can be observed, either empirically or experimentally, and theorized at three complementary levels: macro, meso and microscopic.

The macroeconomic level focuses on the status of the artifact system before and after a given design. The purpose of this exercise is to identify the interoperability constraints that arise before the design and the systemic effects caused by its outcome. Theorizing this level requires us to identify the overall dynamics of a referent artifact system. The macroscopic approach does not allow us to understand the design internally. To do this, more detailed approaches are required. This is precisely the objective of the mesoscopic approach. It aims to describe what happens between the perception of a need and the making of a satisfactory artifact. Design is viewed in this case as a sequence linking well-identified activities and stages. We can define it as a hierarchical and convergent decision-making process (conical model of design). Finally, for an in-depth observation and understanding of design, we have to go down to the microscopic level where the elementary activities carried out by the designer are studied.

PROPOSITION 5.– Since design is both universal and creative, the dynamics of an artifact system (all the artifacts produced by man) are not only gradual[11]. According to artificialists, design alone explains the dynamics of the referent artifact system. Structural effects can certainly come into play, but stating them does not suffice to explain the dynamics. None of these effects will materialize without design and without the designer. Artificialism adds two important points:

– first, the dynamics of a referent artifact system must take into account a sequence of redesigns, called rationalization[12]. This may be due to dissatisfaction with the concept, for example the integration of new constraints. It is, for instance, the will to integrate the concepts of sustainable development in the sustainable city fabric in a better manner in recent years [FOR 11]. However, rationalization can also be initiated because of a questioning of the adequacy of the solutions chosen for the purposes of the artifact. Finally, the dynamics of the referent artifact system can be initiated by the desire for rationalization displayed by such and such group of designers. Its suddenness and rapidity depend not only on structural causes but also on the degree of legitimacy and authority that this group has;

11 The artificialistic perspective differs on this point with the evolutionary perspective on the evolution of systems.

12 For the artificialist, the sequence of re-designs of the artifact or what we call rationalization deserves the same attention as its initial conception.

– second, since design is creative, its outcome can be unpredictable and can provoke jumps that are as brutal as they are sudden within the referent artifact system. Thus, the ambiguity of the laws of evolution of artifacts is highlighted and how a gradual process can bring about the creation of new artifact lines, is specified.

3.5. Interest in thinking about innovation from the artificial perspective

In the preceding pages, we have stressed that we live in an increasingly artificial world and that the artifacts that make up this world owe their existence and their future to design[13]. Design is in fact required when the satisfaction of a need cannot be immediate, by merely taking it from the wild, by making a purchase in a market or by applying solutions inherited from tradition. Hence, for the artificialist, understanding the mode of existence of artifacts therefore involves thinking about the process of design. Artificialism thus appears as a generic theoretical framework of the genesis of innovations[14].

This remark may seem trivial, but it is anything but trivial. Because, if we agree to think of innovation differently, i.e. to consider innovation not from the linear and hierarchical model but from the artificialistic perspective, we will have to identify new possible levers of action.

---

13 With this approach, we agree with the theory of S. Kline and N. Rosenberg who made the design process the backbone of the innovation process.

14 Artificialism has not been developed to serve a discipline (mechanical design, urban space, etc.) or an exclusive set of disciplines: engineering science (SPI – Sciences Pour l'Ingénieur), for instance. It is a generic theoretical framework for the emergence and dynamics of artifacts that mechanics, computer science, biology, economics and management can each mobilize without compromising their own specificity.

3.5.1. *Developing a comprehensive organizational system to ensure the effectiveness and efficiency of the design process*

Adopting the artificialist perspective also impels us to consider the quality of the design process.

In fact, even though design represents on average only 5% of the total cost of a product, it determines 75% of the total cost of the product. Moreover, a wrong definition of the need leads to lots of backward and forward stages that are costly. We can quote the example of the Apple Macintosh, whose development difficulties pushed back the launch, initially scheduled for May 1983 to January 1984. The repercussions of the initial six-month delay (May to July 1983) were estimated. These two months of delay reduced the Mac division's turnover by nearly $10 million and its operating income by $4 million per month of delay [TAR 93, p. 308].

Given its impact on product development costs and time frames, many companies began to improve the cooperation of design partners in the 1980s, so that they could be involved right from the early stages of the process.

Such a perspective challenges the Taylorian type of organization of design, in which:

– research comes up with ideas;

– the design department is responsible for the design of the product;

– the process planning department takes care of designing the production process;

– the production department is in charge of manufacturing the product with the means, tools and ranges defined by the process planning department, etc.

... in favor of a more dynamic and interactive organization more suited to the policy of reducing innovation and design time, such as the “concurrent engineering” type of organization. The latter has the advantage of making it possible to integrate the opinions of the different functions (co-design), industrialization, assembly, etc., in order to avoid a situation wherein design work carried out very competently within each function leads to losing sight of the broader objective of optimizing the entire system [DER 90].

3.5.2. *Thinking of the user*

When considering the design process at the mesoscopic level as a satisfactory problem-solving process, the definition of the need also appears to be a key issue.

3.5.2.1. *Understanding practices and needs*

In fact, designers are unanimous about this: the main problem they come across in their work pertains to the difficulty of pinpointing and identifying the need, in other words, to making sure that the design problem is posed correctly. Experience shows that there are many more product failures and bankruptcies of companies owing to products that are ill-adapted to the market but have been manufactured well, than there are to products that are well-adapted to the market but poorly manufactured.

By way of illustration, we can cite the example of household waste management in the vertical social housing in the Grand Lyon area. The first report cards of the installation of the underground storage silos in Rillieux revealed significant discrepancies between the expectations of the project leaders and those who would have to them (i.e. its users). Although this system is technically operational, it is not operational as far as its use is concerned. The reason for this failure is simple: the designers made the assumption that the user was not a problem (the user is represented as an "*a priori* adherent" of the system), whereas the practices foreseen by the designers have been only partially in line with user needs [BER 15].

We can also quote a better-known example concerning the Apple iPod. Designed in 2001, its sales did not take off because CD player users were not excited enough to buy an Ipod that seems to them to be a useless gadget. It is only after the use of the dematerialized music grew and iTunes downloading platform came up (2003) that the iPod witnessed success that was as unpredictable as it was spectacular (Figure 3.2).

	iPod
2001	125
2002	470
2003	1,451
2004	8,263
2005	31,960
2006	46,432
2007	52,685
2008	55,434
Total	**196,820**

Source: Apple company reports, 2002-2009; eMarketer calculations, February 2009

Figure 3.2. *Apple Ipod unit sales worldwide, 2001–2008 (thousands)*

3.5.2.2. *Functional analysis: an approach turned towards the definition of the need*

Given the criticality of the definition of the need for innovation, a number of design approaches have emerged.

One example is functional analysis, which aims to express the need in terms of expected services rather than in terms of solutions. This assumes that the designer reasons only functionally, without referring to any solution, and proceeds in five stages:

1) Identifying the functions. The objective of the first stage is to carry out an exhaustive review of the functions that the artifact must fulfill.

2) Classification of the functions. The designer classifies the functions by first distinguishing:

– the esteem functions, which reflect the subjective part of the need (for example, the esthetics of an armchair);

– the use functions, which reflect the use of the product, the objective part of the need (e.g. giving support to the body in the case of an armchair)[15];

– the constraint functions that act as limitations of the designer's freedom (regulations, standards and interface requirements), etc.

3) Characterization of the functions according to assessment, levels and flexibility criteria.

4) Hierarchization of the functions.

5) Enhancement of the functions by attributing a weight to them, in relative or absolute value terms.

Although functional analysis is a valuable aid to designers both for design and redesign, there has been a great deal of interest today for what is known as design thinking ever since the publication of P. Rowe's book *Design Thinking* (1987). What exactly does this mean?

15 The difference between esteem functions and use functions can be shown as the difference between a luxury car and a basic small car that each have the same engine. From a use point of view, both cars conduct the same function: they both offer safe, economical travel (use function). The luxury car has, however, a greater esteem function.

3.5.2.3. *Design thinking*

As T. Brown points out, design thinking is not about making an object more attractive, more salable or making it faster than its competitor, but it aims to create a user experience:

> *"Now, however, rather than asking designers to make an already developed idea more attractive to consumers, companies are asking them to create ideas that better meet consumer's needs and desires. The former role is tactical and results in limited value creation, the latter is strategic, and leads to dramatic new forms of value"* [BRO 08, p. 86].

The question that comes up, is how do we go about doing this. According to R. Faste, design thinking appears to be a process consisting of seven stages:

– define;

– study;

– ideate;

– prototype;

– select;

– implement;

– learn.

T. Brown presents it as a process consisting of three stages:

– identify a problem and understand its environment;

– find the concept, the idea that will respond to the problem posed;

– design the shape that will embody this concept.

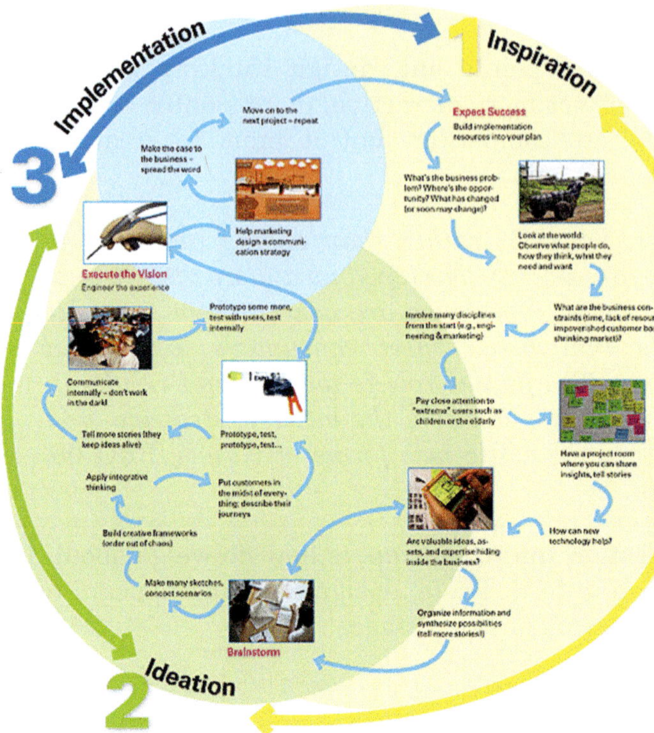

Figure 3.3. *Design thinking according to T. Brown (source: [BRO 08, p. 88–89])*

Although the granularity of the stages appears more or less fine from one author to another, the underlying "philosophy" remains the same.

In fact, design thinking is an intentional and structured process that allows ideas to emerge independently of the size of the problem, the time and the budget available. To be precise, design thinking is:

> *"a methodology that imbues the full spectrum of innovation activities with a human-centered design ethos. By this I mean that innovation is powered by a thorough understanding, through*

direct observation, of what people want and need in their lives and what they like or dislike about the way particular products are made, packaged, marketed, sold, and supported" [BRO 08, p. 86].

The idea is to adopt a human-centered approach and validate the main principles of the solution as soon as possible, according to the "fail fast to succeed sooner" principle. To do this, design thinking favors a certain number of approaches:

– The empathy or experience of the other is thus a key factor in discovering the explicit and implicit needs of individuals and, in doing so, in offering them an appropriate solution. Two tools can then be used: interviews and observation[16], the latter making it possible to see what people are actually doing and what they are not doing;

– Prototype as early as possible. This phase makes it possible to test the ideas as soon as the subject of the thought can be made tangible. The creation of a Lo-Fi prototype also makes it possible to try one's idea very early with future users. At the same time, it makes it possible to validate or revisit the initial assumptions. The design thinking approach is therefore iterative;

– Storytelling of the product concept, using the storytelling method. The idea is to relate the emotions linked to the user experience. Based on the knowledge gained from the interviews and observations, it is possible to start from the current situation and move on to the future experience. This storytelling, which can take varied forms such as storyboard, video etc., makes it easier to correct the design.

Far from wanting to minimize the contribution of design thinking to the "fabrication of innovation", we must emphasize that this human-centered approach is obscuring

---

16 We will discuss the issue of observation at greater length in Chapter 5.

the issue of the specificity of rationality at work[17]. However, as we shall see subsequently, our not considering how creative rationality unfolds theoretically masks the fact that the issue of value creation (or desirability for the user), the issue of technical feasibility and the issue of economic viability will only come up after the emergence of creative ideas.

Let us make no mistake about what we are trying to say here. We are not saying that the deployment of creative rationality will not lead to value creation, but we are just trying to humbly recall that in order to create the said value, we should be in a position to deploy this form of thought. Therefore, in the following chapter, we will focus more specifically on sketching the outlines of this form of thought.

17 Various reasons can be cited to explain this. The first consists of saying that perhaps we still cherish the idea of a good savage (or a disheveled bearded man in his garage) who has dazzling intuitions, a romantic vision of a superman with the capabilities of a visionary genius. The second reason may stem from the fact that the narratives of great innovators favor the outcome more than the rationality at work.

4

Innovating by Implementing Creative Rationality

In the previous chapters, we have focused on presenting the artificialist perspective and outlining the contribution of such a viewpoint to the management of innovation. However, considering innovation based on the artificialist perspective also enables innovation to be conceived from the viewpoint of the rationality at work.

This idea is not new, and several authors had already underlined the creative nature of the design process: "*design involves (...) the presence of a creative step*" [ARC 84, p. 58], "*all designing is iterative, using creativity and compromise to move from a field of possibilities to one unique solution*" [ROY 86]. Their viewpoint thus falls in line with E. Phelps' theory, according to which innovation derives mostly from the imagination of men and women from all backgrounds. In the 19th Century, innovation was open to everyone according to Phelps, and any tradesman or worker could discover something. The inventors of the industrial revolution were neither scientists nor people with advanced training, as E. Phelps writes: "*Watt was the exception, not the rule. Arkwright was a wigmaker turned industrialist, not a scientist or engineer. Hargreaves, a Lancashire weaver (...). The great Stephenson was virtually illiterate*"

[PHE 13, p. 12]. What they had in common is that they were inventive and creative individuals engaged in a trial-and-error process. We also agree with the argument of R. Florida, who points out that "*Human creativity is the ultimate economic resource. The ability to come up with new ideas and better ways of doing things is ultimately what raises productivity and thus living standards*" [FLO 02, p. xiii].

By accepting that the action of a designer is creative means that the action in question leads to a result and follows a process that is simultaneously original, ingenious, and unthinkable *a priori* as designing, involved in situations, is never determined by them. It follows its own logic. We are then led to discard the analysis of the design process in favor of that of the rationality at work. Thus, we are encouraged to take creative rationality into consideration.

4.1. Creative rationality: what exactly are we talking about?

4.1.1. *Thinking in terms of relation*

Considering the creative nature of the design process leads us to update a form of thought, long excluded from the domain of science, which the Greeks called *mètis* [FAU 12][1].

As M. Détienne and J-P. Vernant [DET 74] underlined, since the 5th Century, *mètis* has been discarded as non-knowledge and non-thought by philosophy for reasons related to both its practices and in the name of a metaphysics of being and the unchanging. According to

1 In our previous works, we hypothesized that technology may be conceived as the science of creative rationality [FAU 12]. Considering technology as the science of creative rationality allows us to open a tradition of thought that does not merely try to account for the effects of technology or its ethical implications.

H. Vérin, however, *mètis* is rationality implemented by engineers [VÉR 93, p. 16][2].

Mètis is similar to the inventive form of thought that G. Vico theorized as *Ingenium*. In his *De studiorum ratione*, G. Vico presents ingenium as the faculty of discerning relationships among things. It is the ability of designers to bring together apparently distinct worlds (importing furnishing materials into the design of glasses), to find a link where there was none, to find out what happens outside one's field, and to be open to anything. It is the faculty used by the inventor of the Klymit jacket, N. Alder, who incorporated what he learnt about argon (isolation) during a diving training course into the design of snowboard jackets to retain heat [CHR 13, p. 134].

Ingenium appears then as an idea of relation: "*As Poincaré noted 'To create consists of making new combinations of associative elements which are useful' (p. 286). Creative ideas, he further remarked, 'reveal to us unsuspected kindships between other facts well known but wrongly believed to be strangers to one another' (p. 115)*" [MAR 05, p. 137]. It is thus the source of innovation: "*I think that this ability to combine different types of knowledge, to go look here and there for ideas that are unrelated to our own field, represents the first rule of innovation (...)*". [JAC 94, pp. 46 47]

This remark may seem trivial unless, by following G. Vico, we put it into perspective with the ability of analytic reason to produce innovation:

2 Sophocles, in the first *stasimon* of Antigone sung by the chorus of Theban old men (a famous excerpt known as the Praise of Man) focuses on the "*ingenious man*". His *mètis* is what distinguishes him from the other animals and makes him actually superior. "*Wonders are many, and none is more wonderful than man*".

> *"None of the great technological inventions that have changed the global landscape is, according to the* De ratione, *the product of the analytic method recommended by Descartes, and most of them come before the development of mathematical physics, be it cannons, sailing ships, clocks, or Brunelleschi's cupola. In a marginal note of the* Scienza nuova, *which was not published in the definitive edition of 1744, Vico went as far as saying that in the Middle Ages, in the 'barbarous times come again', 'all the greatest inventions were made by ignorant or barbarous men'. For example, the nautical compass was invented by a shepherd from Amalfi, the telescope was invented by an uneducated Dutch spectacle-maker, and, according to Marco Polo, 'blood circulation and printing were invented in Great Tartary' (S.N., section 1246)*". [PON 03][3]

Thus, ingenium is similar to the idea of bisociation such as it was defined by A. Koestler. The concept of bisociation underscores how creativity does not emerge *ex nihilo*: "*it uncovers, selects, reshuffles, combines, synthesizes already existing facts, ideas, faculties, (and) skills*" [KOE 64, p. 120]. Besides, it mostly makes it possible to move away from the logic of association (which, according to A. Koestler, involves thinking on only a single plane) by highlighting a confluence of elements that belong to two distant universes, leading to a

3 Let us make no mistake about the meaning of our remark. Associating innovation with the use of creative rationality does not mean discarding analytic rationality. These two forms of thought are in fact mobilized during the design process, but their contribution differs. Besides, let us underline that recent research carried out by J. Kounios, the director of the Creativity Research lab, and M. Beeman, seems to confirm, based on the study of brain activity, the existence of these two forms of thought. "*People who are highly analytical have more activity in their left frontal lobe. And people who are highly insightful have more activity in their right posterior parietal lobe, in the back of the brain*" [SCH 15].

creative synthesis thanks to the "*underground gambling of the mind*".

As A. Koestler points out, the creation of the concept of bisociation aims to distinguish between a routine form of reasoning, which he regards as taking shape on only a single plane, and the creative act, which always operates on several planes and "*connects previously unconnected matrices of experience*". [KOE 64, p. 36].

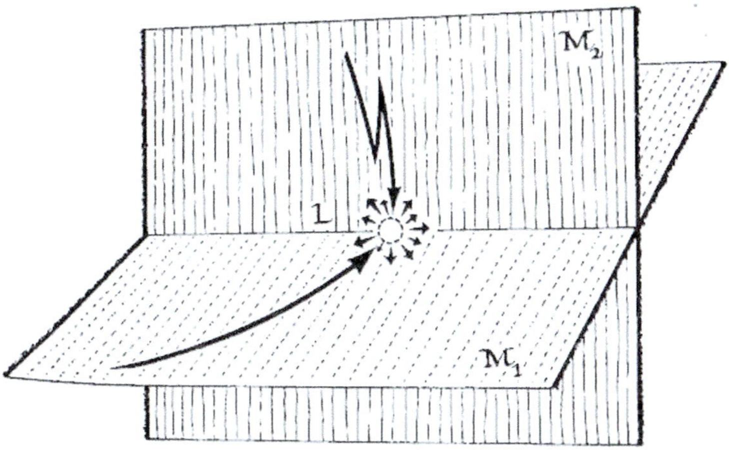

Figure 4.1. *Bisociation according to A. Koestler (source A. Koestler, 1964: 36)*

We come across what F. Johanssen had called the Medici effect while referring to the creative explosion that marked Florence during the Italian Renaissance.

4.1.2. *A form of thought that can replace the inexplicable with the rational*

Ingenium refers to what we have called "creative rationality" since, as A. Pons points out, no French term can truly convey the meaning of the word *ingenium*.

We are aware that the link established between the two terms "rationality" and "creativity" may seem somewhat oxymoronic, given how often we regard rationality as opposed to creativity.

As D. Burkus underlines in his work *The Myths of Creativity*, as their knowledge was not enough to account for phenomena, the ancient Greeks developed myths: "*The myths developed were an attempt to explain the mysteries they (culture) couldn't really understand*" [BUR 14, p. 1]. In other words, myths offer a readily available explanation of the organization of the world. The myth of divine intervention is precisely one of these myths. Thus, in the *Ion*, one of Plato's dialogs, Plato points out that poets are nothing more than interpreters of the gods: "*Plato argued that a poet is able to create only that which the Muse dictates (…), Rudyard Kipling (1937/1985) referred to the Daemon that lives in the writer's pen (…). Many people seem to believe, as they do about love (…) that creativity is something that just doesn't lend itself to scientific study, because it is a spiritual process*" [ALB 05, p. 5][4].

The idea of creative rationality, however, agrees with a way of thinking that breaks with the myth of a creation resulting from divine inspiration or exceptional abilities specific to rare geniuses who can imagine the whole world intuitively, without following any of the canons of reason[5].

4 In the 19th Century, Muses still governed poetic creation: V. Hugo and C. Baudelaire frequently invoked them in their works.

5 Creativity is not related to a genetic characteristic. This hypothesis was confirmed by the study carried out by M. Reznikoff, G. Domino, C. Bridges, and M. Honeymon [REZ 73], who analyzed the creative abilities of 117 pairs of twins aged 13–19 who had to take a series of 10 creativity tests. Their research led to the conclusion that "*The overall results, however, failed to provide convincing evidence of a genetic component in creativity*".

Considering creative rationality leads us to perceive creativity in terms of process and, therefore, to replace creation with creativity as well as the inexplicable with the rational [FAU 07]. If R. Lester and M. Piore [LES 04] present this process as an interpretative process involving conversations capable of bringing about some possibilities, A. Hatchuel, P. Le Masson, and B. Weil [HAT 07] claim that creativity is intelligible if we consider the C-K theory. According to the latter, it is possible to theorize the emergence of novelty based on a theory of design reasoning centered on the distinction between two types of space: the C space, which represents the space of concepts, and the K space, which represents the space of knowledge. They point out that, based on a given problem, design reasoning begins with a C-K conjunction that transforms a concept into knowledge. The C-K theory reaffirms thus the role of knowledge in innovation. Without knowledge, no ingenious combination can take place, as design reasoning operates based on the knowledge included in the C space. The main contribution of the C-K theory, however, lies in the emphasis put on the production of knowledge at the very source of innovation since, as the authors point out, without concept we are doomed to explore indefinitely objects whose definition never changes.

As we have understood, linking the two terms "rationality" and "creativity" means underlining that creativity also involves rationality, that it is not an irrational process, and that it can be modeled and taught as such in engineering schools (we will come back to this point later). It also means reaffirming that each of us is creative[6]. Thus, we

6 It means breaking with the widespread belief that "not everyone can be creative", a categorical belief according to which some are "chosen" individuals bound to produce ideas, while others are doomed to lack ideas and follow those of others.

agree with the conclusions of J. P. Guilford who, in an article called "Creativity" published in 1950, already pointed out that only the layman thinks that the creative individual has a unique gift that the common man lacks. He specified that such a way of conceiving creativity was rejected by all psychologists, according to whom every individual has a certain amount of creative abilities [GUI 73].

4.2. The reality of creative rationality

Since we are focusing on the creation of artifacts based on the artificialist perspective, we can rely on what we will call innovation biographies in order to show how this way of thinking actually works.

4.2.1. *What are innovation biographies?*

The biographic approach is most often chosen to show how some individuals managed to make great discoveries or invent great things. Therefore, biography is a type of writing that focuses on an individual, but disregards the innovation process. This is why we intend to leave aside the biography of innovators in favor of the biography of innovation.

We are aware that such an approach may make readers stop and think. Etymologically speaking, biography comes from the Ancient Greek words βίος ("life") and γραφή ("writing"). Therefore, the term "biography" refers to a type of writing that focuses on the story of a specific life. Thus, referring to innovation, biography means thinking that artifacts have a life. As T. Bonnot underlines [BON 02], if we understand the idea of life from a purely biological point of view and we regard it as limited to human beings only, our reasoning is guided by an excessive purism. According to him, ever since an object is transformed (technologically, physically or even routinely), we can say that it has a life. Therefore, he claims that the term "biography" can be applied to objects. If they are transformed, they have a life; if they have a life, we can consequently narrate it and write their biography [BON 02].

Considering the mode of existence of artifacts, based on innovation biographies, leads us to consider the context (economic, technological, scientific, etc.) of the emergence of said artifacts and take into account the multiplicity of the stakeholders, their roles, their projects, and their representations and imagination. Therefore, innovation biographies show us how artifacts convey meaning, human values, and how they are related to structures and choices. Solar Impulse (the solar airplane that can fly day and night with no fuel) is an emblematic example. According to B. Piccard, the Solar Impulse project[7] aims to demonstrate the effectiveness of new clean technologies to protect the natural resources of the planet. More than a tour around the world, Solar Impulse is a plea for the environment and the proof that clean technologies can achieve the impossible[8]. Solar Impulse is more precisely the ambassador of a vision where it is easier to protect the environment by implementing clean technologies en masse rather than fighting the societal tendencies of mobility, comfort and growth.

Thus, innovation biographies lead us to discard the idea of an alienating technology that enslaves man. However, as they are not centered on the life of an individual but focus on the life of an artifact, innovation biographies can also detect, as we will see later on, the use of creative rationality.

---

7 We will soon come back to this example.

8 "*The future is clean*" is what B. Piccard declared to the applause of the crowd upon arriving in Abu Dhabi.

Figure 4.2. *The meaning of innovation (source: https://twitter.com/bertrandpiccard). For a color version of the figure, see www.iste.co.uk/forest/innovation.zip*

4.2.2. *The example of Gutenberg's printing press*

If we start writing the biography of Gutenberg's printing press, we realize very quickly that the printing press resulted from an ingenious combination of knowledge.

The first books were rare, as they were copied one by one by hand (hence the name of "manuscript"). This work, carried out especially by copyist monks in a room called *scriptorium*, was laborious (sometimes it took more than 10 years to rewrite the Bible by hand) and very expensive. From the 13th Century onwards, the demand for books had increased and if copying workshops were to be created, this system would have suffered from low productivity and books would have remained too expensive for most readers. Understanding the creation of the printing press also involves taking into consideration a specific technical context:

– a wood-engraving process called xylography was already used[9];

– Nuremberg, which is near Mayence, was famous for its precision metallurgy;

– similarly, artillery developed from the 15th Century onwards: "*bronze alloy, mold? Is this not the same technological flow of the printing press?*" [BEC 92, p. 50];

– the press, the tool that gave its name to impression and printing, was already used to press grapes;

– the production of hemp and linen paper was growing in Europe;

– India ink had been used for a long time in the Far East and in the Middle East.

A cursory analysis of the situation may lead us to the conclusion that the emergence of the printing press was bound to happen, as in the end, all the ingredients were there. As G. Bechtel underlines: "*these conditions have played such a marginal role for the printing press that most of them had already been in place for about fifty years: paper does not date back to 1450...*", besides pointing out that studying the circumstances explains more the conditions favorable for the inventions than their emergence [BEC 92, p. 43][10].

9 Xylography involved engraving the text of the page to be printed in wood. This process had several drawbacks: it was impossible to modify the text at a later stage, and it was difficult to produce characters with a regular shape, without forgetting that further impressions damaged the wood.

10 Let us also underline that the works mentioning, following on from A. Marshall, the fact that innovation was in the air or describing cultural breeding grounds "*Leonardo da Vinci lived during the Italian Renaissance, together with Michelangelo, Raphael, and many other Renaissance engineers and scientists*" [DOR 15, p. 11] helped decenter "*the event-driven history of creative geniuses towards the innovation milieu*" [DOR 15, p. 11].

The "stroke of genius" in the process conceived by Gutenberg (1400–1468) was the development of metal mobile characters and his ability to make good use of the techniques used for other purposes (the press) and improve them. We can mention here the example of the ink used for printing, which he made thicker than the India ink used until then in the Far East and the Middle East to avoid smudges.

In the end, Gutenberg's genius lay in his ability to bring together pre-existing techniques: "*Gutenberg's invention is strictly a bringing together of many inventions and the modified application of known working practices from printing to type*" [KAP 96] or, in other words, in his ability to deploy his creative rationality.

4.2.3. *The example of the printing press is not an isolated case*

This knowledge crossing can also be found in the emergence of the Lumière brothers' invention of the cinematograph in 1894. M. Faucheux's work *Auguste et Louis Lumière* [FAU 11] clearly shows that the invention of the cinematograph lies at the convergence of the growing photographic industry (the famous blue-label plates produced by the Lumière de Monplaisir factory), the chemical industry and the textile industry, as the invention of the cinematograph borrowed the "presser foot" mechanism from sewing machines. Thus, in a cinematograph, a film is immobilized for a short instant at each image. Thanks to this ingenious method, it is possible to flip through different and yet very similar images, decomposing a movement at a speed high enough to recreate the movement and give the illusion of a continuous image.

This knowledge crossing can finally be found in the medical field, as is shown by the emergence of vascular surgery and the innovative technique of triangulation suturing conceived by A. Carrel. Carrel's project was to carry out vascular anastomoses without causing any stenosis or thrombosis due to the constriction of the sutured vessels. In June 1902, he reached the conclusion that the only feasible technique was suturing, i.e. "sewing vessels". This conclusion was related to the fact that A. Carrel's mother was a lace maker in Lyon and that he learnt increasingly finer knots from a famous lace maker in Lyon, Mme Leroudier, who was a friend of his mother [FAB 12].

If the previous examples confirm the role of knowledge crossing or the use of creative rationality, innovation still raises the issue of the scope of this remark. Are we witnessing the same phenomenon today?

We may suppose that the fact that Virgin Atlantic has managed to stand out from the competition because of the entertainment services and the experience it offers to passengers on its transatlantic flights may be related to the fact that its founder, Sir R. Branson, comes from a musical background. Thus, he injected into the airline industry the "hip" and "cool" values associated with the record label Virgin Records.

In line with C. Christensen, J. Dyer, and H. Gregersen [CHR 13] and their work *The Innovator's Genius*, it seems that the answer is also positive. These authors explain how making unprecedented and unlikely associations, with the potential to lead to actually revolutionary projects, is precisely a characteristic of innovators. Moreover, they point out:

– that innovators have rarely innovated *ex nihilo*, but that they reorganize existing ideas;

– that what they had in common was "*to love collecting the largest possible number of ideas, just like children love collecting Lego blocks*" [CHR 13, p. 42].

C. Villani seems to confirm these authors' theory. While mentioning the birth of the theorem that led him to win a Fields medal, he points out: "*In my own research, the most striking aspect lies in associating apparently unrelated subjects. This sometimes results in stunning ricochets (...) which, adjusted in relation to the rest, miraculously give us the solution to a conundrum we have been stumbling over for months*" [VIL 14].

4.2.4. *Towards an adventurous transgression*

The aforementioned cases confirm the idea that creative rationality encourages knowledge crossing. However, this crossing has to do with what we will call an adventurous transgression.

What do we mean by adventurous transgression?

A crossing approach and the combination of types of knowledge belonging to different universes lead us away from the established norms and paradigms. In this respect, we agree with A. Koestler who, in his work *The Act of Creation* [KOE 64], underlined that when we are dealing with a problem we have come across before, we solve it using solutions that have already been tested. These solutions become routines that must be disrupted in order to be creative: "*The act of discovery has a disruptive and a constructive aspect. It must disrupt rigid patterns of mental organization to achieve the new synthesis*" [KOE 64, p. 104].

This is what C. Villani quickly realized when he conceived his theorem: *"deep down I am convinced that the solution requires completely new tools, and must allow us to look at the problem from a new angle. I need a new norm"* [VIL 12, p. 48]. We can clearly see this need to free oneself from established paradigms when we reconsider the history of the creation of Swatch, the famous Swiss watch, and Seb's Actifry fryer.

Until the beginning of the 1980s, Switzerland was famous for its expert watchmaking and its ability to design ultra-high precision watches. The emergence of low-cost quartz watches from Asia, however, led to a dire crisis in watchmaking in the 1970s, to the extent that the Swiss watchmaking industry was considered to be doomed at the beginning of the 1980s. It is, in this context, that E. Mock and J. Muller conceived a watch that could be mass produced at low cost. At the beginning of the project, E. Mock and J. Muller were completely alone. *"No one wants to work with them. No one wants to run the risk of getting involved with 'two crazy men who will screw up (...) we have never made the same watch a million times; no one will want to buy a product like that' is what people in the business said"* [GAR 12, pp. 35–38]. They conceived a watch that was not regarded as such in the eyes of the "Swiss tradition": a watch that contained a plastic-molded case and a glass (also plastic-molded) fixed on the watch glass by ultrasonic welding. These choices, made in order to obtain an inexpensive watch, ended up making Swatch a watch that could not be repaired, *"whose identity broke with the traditional Swiss watch that lasts over time"* [GAR 12, p. 49].

We may also mention the example of Seb's Actifry fryer. At the beginning of the 2000s, mass distribution flooded the market with low-cost smaller and average household appliances made in China. Seb was violently struck by the effects of globalization and saw its market of electric fryers shrink: "*Between 2000 and 2005, the number of units produced was to collapse from 2 million down to 500 000, as pointed out by Thierry Coutureau, the head of research on electric cooking appliances*" [LUP 13]. The factory in Is-sur-Tille (Côte-d'Or, France), which produced classic fryers, was then threatened with closure. It is, in this context, that the idea of making tasty but non-fatty fries was born. At the beginning of 2007, Seb launched Actifry, which can cook a kilo of fries with only a spoonful of oil and costs 200 euros. At the time, not many people believed in this innovation given how dramatically it broke with the traditional idea of frying. However, Actifry was to become incredibly successful. The group hoped to sell 17 000 fryers in the first year. It eventually sold more than 10 times that figure. Today, it sells its products in 57 countries.

However, freeing oneself from established norms and paradigms is not as simple as it may seem at first sight.

C. Carrier, for example, underlined that when an expert in a given field comes across an idea that has just disrupted the established theories[11], he must be very brave and determined to accept attaching enough importance to it to keep exploring it, pointing out that "*he has a lot to lose and, in order to do this, he will have to give up on some ideas that have always sustained and guided him*" [CAR 97, p. 17–20]. Therefore, expertise may slow down innovation. This is what has been suggested by B. Piccard, who points out that creativity and innovation do not come from within the system, as the system is too fossilized by the *a priori* to be able to invent something new. He mentions, as an example, that it was not car manufacturers who created the

11 Expertise provides a simplistic approach of reality by drawing the boundaries of what is possible and what seems impossible. A. Einstein went even further, saying that "*If at first the idea is not absurd, then there is no hope for it*".

best electric car. It was a billionaire who made a fortune with the Internet who conceived Tesla [PIC 14][12].

Similarly, since transgression is always, at least temporarily, against the established order, even if it often ends up involved in another concept of the order [ALT 00], it clashes against the prevailing conservatism[13].

Could we find any better illustration of this conservatism than the reception of A. Carrel's works in 1902?

As a candidate to hospitals, he was only met with irony and hesitation, and was treated similarly by J.-P. Morat, a physiology professor who, irritated by the rumor about A. Carrel's performances, published an article where he warned against illusions and recalled that transplants had no future as long as the vasomotors were not linked themselves [SOU 74]. In 1903, subject to the ostracism of his professional entourage, A. Carrel emigrated to Canada and then to the United States, where S. Flexner, the director of the new Rockefeller Institute, offered him a fellowship in 1905. However, in 1912, he obtained the Nobel Prize in Physiology or Medicine *"in recognition of his work on*

12 This observation leads us to discard the idea that the use of creative rationality is the prerogative of experts. Thus, we agree with the conclusion reached by B. Jacomy who, based on the example of R. Moreno's invention *Take the money and run* (better known as chip card), already underlined that *"those who break with tradition and invent the most amazing things often are not experts"* [JAC 94, p. 46]. The history of technology contains many examples in support of B. Jacomy's theory. It was, for example, a painter, S. Morse, who invented the electric telegraph; a joiner, Z. Gramme, invented the dynamo in 1868; a carpenter and millwright, L. A. Pelton, invented the turbine in 1879; and a journalist, L Biró, invented the ballpoint pen in 1938.

13 By conservatism, we mean not an ideology but a tendency to see things and adopt a specific stance, such as the tendency to choose what we know over the unknown or (actual) gain to potential gain.

vascular suture and the transplantation of blood vessels and organs"[14].

On an industrial level, several examples of the expression of this conservatism can also be found in the course of history. We may mention, as an example, the story of G. Stephenson. As S. Smiles [SMI 12] underlines, when G. Stephenson tackled the task of developing a new, quicker and safer traction mode than horses, which would give rise in 1930 to the first steam locomotive able to carry at once passengers and goods, he met with considerable hostility: "*Railways had thus, like most other great social improvements, to force their way against the fierce antagonism of united ignorance and prejudice*" [SMI 12, p. 338]. We can also find the trace of the expression of this conservatism during the emergence of the Internet. Although L. Pouzin created, in 1971, the data packet exchange protocol at the source of the Internet, at the time the Internet seemed a gimmick for researchers with no interest from the viewpoint of political decision makers. This is the reason why V. Cerf, an American, would later become the official father of the Internet. It is exactly this inertia that led some of the researchers working for Xerox to leave Xerox at the beginning of the 1970s for other companies, where their ideas were rapidly adopted: laser printing by HP, the Ethernet protocol by 3com and the graphical interface by

14 We should point out that I. P. Semmelweis, a Hungarian obstetrician, received the same treatment when he showed the benefit of washing hands after the dissection of a corpse and before delivering an infant. "*Mr Semmelweis claims that we carry on our hands small things that would be the cause of puerperal fever. What are these small things, these particles that no eye can see? It is ridiculous! The small things of Mr Semmelweis exist only in his imagination!*" said Klin, the head obstetrician in Vienna who dismissed I. P. Semmelweis [LAB 09, p. 90]. In line with what A. Lumière says in his work called *Les fossoyeurs du progrès. Les mandarins contre les pionniers de la science (The gravediggers of progress: mandarins against the pioneers of science)*, when we analyze the history of science, we realize that geniuses often met with the lack of understanding or bad faith of their contemporaries.

Apple [FRE 14]. This also explains why the AZERTY keyboard was kept in France[15].

Besides, if we say that this transgression is adventurous, it is because it contains all the ingredients of an adventure. Using one's creative rationality leads us to explore territories where no one has dared to venture as, to use G. Garrel and E. Mock's phrase, "*We innovate by taking back roads, which are deserted, since we pick mushrooms on small paths rather than on motorways*" [GAR 12, p. 1]. If we take such paths, we rapidly come to face twists and turns that, to be negotiated, require not only solid motivation but also perseverance since, as C. Villani points out when recounting the creation of his theorem, how "*could I have imagined that it was so difficult, I've never done this*" [VIL 02, p. 82], "*Version 55 (…) a new hole has appeared. I'm railing. Enough is enough!*" [VIL 12, p. 151].

Taking back roads means accepting wandering like Ulysses, who can grasp the concrete diversity of reality and face the unexpected to return to Ithaca[16], or accepting that we will "change altitude", to use B. Piccard's expression (2014). This adventurous transgression was especially prominent in the creation of Solar Impulse 2.

---

15 Computer keyboards have been copied from the arrangement of letters in typewriters, which had been conceived due to mechanical constraints, i.e. slowing down typing to prevent letters from becoming stuck. This arrangement had not been thought for French. Based on the study of the frequency of letters in 400 000 French words, C. Marsan put forward, in 1976, a new design that separated keys into two groups, so as to speed up typing and make it more user-friendly. Despite being designed specifically for the input of French words and regardless of an AFNOR norm established in 1987, C. Marsan's keyboard did not become widespread, explaining why the AZERTY keyboard remains standard in France, even on smartphones.

16 In the Odyssey, Ulysses is called "polymetis". He is the master of "metis" and creative rationality, who uses his ingeniousness to return to Ithaca on an adventurous crossing that takes him from island to island.

4.2.5. *The Solar Impulse project*

B. Piccard[17] came from a family of explorers. His grandfather, a physicist called Auguste Piccard, designed the first pressurized capsule that allowed him to become in 1931 the first man to reach the stratosphere. His father, an oceanographer called Jacques Piccard, spent his life designing machines to explore the oceans. In 1960, he adapted his father's capsule to dive down to 10 916 meters below the sea level. B. Piccard, a doctor-psychiatrist and a fan of aeronautics and aviation, took up challenges that were regarded as impossible. He completed the first non-stop tour around the world in a balloon. After two failed attempts, he ended his tour around the world with B. Jones on the 21st of March 1999 in 19 days, 21 hours and 47 minutes. It was after this tour around the world in a balloon that the idea of Solar Impulse was born. When he landed in the Egyptian part of the Libyan desert with his co-pilot B. Jones after 45 755 kilometers covered in the *Breitling Orbiter 3* balloon, there was no more than 40 kg of propane left, out of the 3.7 tons loaded before take-off in Château-d'Œx: "*We had barely enough fuel to finish the journey, which is why I vowed that during my next adventure I would free myself from this fuel-dependency. It started with my vision of a solar-powered airplane flying around the world with perpetual endurance*" [PIC 16]. This dependence on fossil fuel became especially evident in the final phase of the mission, when B. Piccard vowed that he would travel around the world a second time without any fuel or harmful emissions.

In 2003, when he began to discuss with A. Borschberg[18] a plane with the wingspan of an Airbus A340 which could fly

17 B. Piccard resembles the hero of a modern version of a novel by Jules Verne, mixing futuristic technologies in the service of one cause: flying without fuel or pollution.

18 An engineer, with a degree in management science, trained as a fighter and commercial plane and helicopter pilot.

non-stop powered only by solar energy, every aviation expert told both of them that it was impossible.

It seemed impossible since, in order to fly day and night powered only by solar energy, Solar Impulse should have unmatched performances in terms of aerodynamic and energy efficiency. However, this project also seemed unfeasible in that it involved a paradigm shift: discarding the concept of a means of transport that burns fuel to operate in favor of a means of transport where energy reserves increase during the flight.

B. Piccard and A. Borschberg had to work with no point of reference whatsoever and gave their all, using ultralight materials and new types of construction. This requirement led B. Piccard and A. Borschberg to surround themselves with a network of actors with high-level knowledge. We refer specifically to:

– The École Polytechnique Fédérale de Lausanne (EPFL), which contributed its knowledge in terms of ultralight structures, energy-chain management and man–machine interfaces;

– The European Space Agency (ESA), which provided knowledge in several cutting-edge technological fields: solar batteries and cells, energy management systems and ultralight building materials;

– Dassault Aviation, which contributed its expertise both in the design of the solar plane and in fields such as aeroelasticity and flight control, as well as system safety and reliability;

– Omega, which could optimize all the energy chains of the prototype in temperature conditions ranging from –40 °C to +55 °C and developed an on-board tool as revolutionary as the plane which, indicating to the pilot the bank angle of the airplane to the nearest degree, would provide him essential assistance when flying as well as landing.

The combination of this knowledge resulted[19], after years of research and development, in Solar Impulse 2, a one-seater carbon fiber aircraft with a wingspan of 72.30 m[20], weighing 2300 kg[21], and reaching a speed of 90 km/h during the day and 60 km/h at night[22]. This plane managed to finish a historic tour around the world by landing on the 26th of July 2016 in Abu Dhabi, the capital of the United Arab Emirates, at 04:05 local time, after leaving from there on the 9th of March 2015 and traveling 17 legs across Asia, the Pacific Ocean, the United States, the Atlantic Ocean, the Mediterranean and the Middle East, i.e. a journey of nearly 40 000 km without a single drop of fuel.

As we have seen, Solar Impulse 2 was born because B. Piccard and A. Borschberg did not hesitate to cross the boundaries and revolutionized knowledge mapping. If the names of Auguste, Jacques and Bertrand Piccard are now associated with great inventions, it is because all three embarked on an adventure designing artifacts and achieving what no one before them had thought possible. B. Piccard pointed out in a tweet sent on the 13th of September 2016 that "*creativity is not having one more idea, but one less certainty*".

---

19 We should note that we can discern, in this example, the traces of the use of creative rationality, i.e. a synthesis of knowledge at the very source of innovation.

20 That is, the wingspan of an Airbus A380, in order to minimize induced drag and offer the largest surface possible for the solar cells.

21 That is, the weight of a car after paying attention to each gram in order to make an ultralight plane to reduce its energy consumption.

22 That is, the power of a scooter, after the maximum optimization of all the energy chain.

4.2.6. *A journey to the center of the production of knowledge*

In the previous sections, we have seen that innovation results from the use of creative rationality which, by combining knowledge, produces new knowledge at the very source of innovation. However, we face the issue of sketching the outlines of a model of intelligibility of the production of knowledge associated with the concept of creative rationality. After presenting a basic model, we will point out the reasons that explain the limited nature of this development. Finally, we will broaden the basic model by considering the issue of the production of knowledge beyond the reasoning used.

4.2.7. *The basis of a creative rationality model*

At the beginning of our model is the idea that without knowledge an ingenious combination cannot occur and that without a problem creative rationality cannot start.

The knowledge production, noted I, resulting from the creative rationality of an actor, is a function of the knowledge he mobilized (Km). Let us assume that there exists a relationship between Km and, on the one hand, the current state of knowledge in society (Ks), and on the other hand, his initial knowledge acquired during his studies and those acquired by the production of knowledge through experience[23].

23 An individual's knowledge is not an immutable entity. It must be considered dynamically, all the more so as there is a phenomenon of erosion and devaluation of knowledge acquired through experience. Due to the development of new knowledge and technologies, some knowledge is actually becoming obsolete.

The scope for the expansion may be greatest when varied pieces of knowledge are brought together. The variety of mobilized knowledge effectively determines the more or less innovative level of the combination. In the recent literature, there is an increasing consensus that resource heterogeneity provides a clear potential for learning and innovation [NOO 07]. If we have access to the same piece of knowledge, we will investigate the concepts whose definition never changes "*Little progress would be made in a world of clones*" [MAS 01, p. 220]. This idea has been confirmed by C. Christensen, J. Dyer, and H. Gregersen [CHR 13], who clearly showed why, by increasing our stock of knowledge, we multiply the number of possible combinations and, therefore, the potential for innovation.

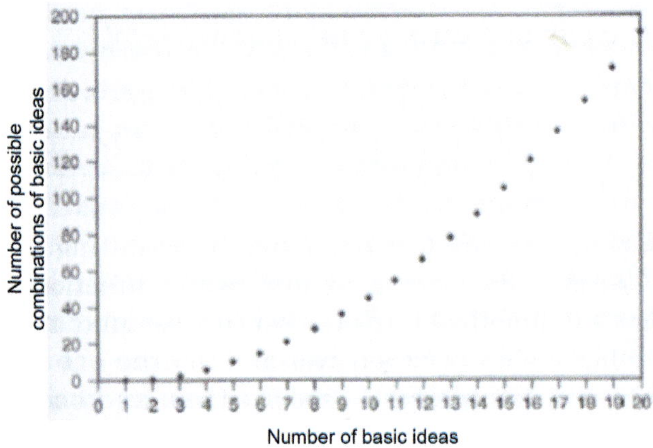

Figure 4.3. *The relation between the number of ideas and the number of possible combinations (source: [CHR 13, p. 43])*

Theoretically, the combination of knowledge need not be a collective activity. However, in current design situations, such an activity is required. Our contemporary knowledge is specialized[24] and complex product development requires the cooperation of different actors, skills, knowledge domains, sciences, technologies, etc. Experience shows that it is very difficult for an actor to master several disciplines. Thus, I is a function of knowledge held by the different actors implied in the process.

The actor population being noted as A = {1, …, n} at a given time t, we can obtain the production function:

$I = f(Kmit).Kst$, with $i = 1,\ldots,n$

Two points must be specified.

First, each actor $i \in A$ is endowed with a vector of knowledge. However, the potential of the interaction is not a simple addition of the knowledge these vectors contain. We must take into account the actual complementarity of the knowledge vectors. Some knowledge can indeed be substitutable. Besides, the design project requires, for every actor, only a part of his knowledge vector.

Moreover, if the variety offers potentialities and opportunities for interesting expansion, the latter can occur only if the cognitive distance [NOO 00] is not too great. Nooteboom points out an interesting idea. The "cognitive distance" provides an opportunity to learn from others who, according to their own experience, interpret, understand and estimate the world differently. If the cognitive distance is increasing, then opportunities of new combinations favorable to innovation will appear. Beyond a certain threshold, the cognitive distance becomes too great and leads directly to a common misunderstanding.

---

24 According to A. Smith, knowledge specialization is the logical outcome of the division of labor and the prerequisite for further knowledge creation [MAS 01].

4.2.8. *A limited production of knowledge*

The absorption capacity of a design actor is indeed a decreasing function. The more the cognitive distance increases, the more it is difficult to understand each other. The lack of a common language, shared values or shared perception induces a cognitive cost. The interaction can fail.

Innovation then depends on a parabolic function of the cognitive distance in the form of an inverted U [NOO 07]. Figure 4.4 represents the graph of this function. The summit of the parabola represents the optimal cognitive distance between the actors. This optimal distance is the distance:

– which is wide enough to allow innovation;

– weak enough to allow effective cooperation.

At this point, many efforts would be required to overlap ambiguities and eliminate mutual misunderstandings.

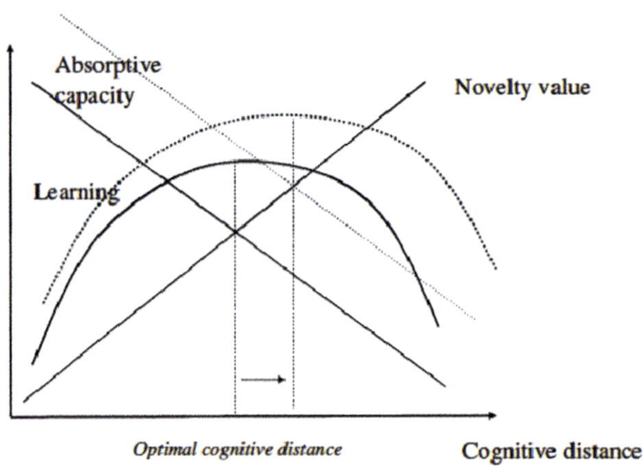

Figure 4.4. *Optimal cognitive distance (according to [NOO 01])*

The main result we can take into account by applying Nooteboom's model is that creative rationality is driven by the bounded rationality principles and by the cognitive distance between design actors. We can add that their personality, their respective experience or expertise must be integrated in a unified model of creative rationality. Several studies based on creative actors have shown that they master their domain to overcome it. According to J.R. Hayes (1989), an average period of 10 years is required to master a domain Figure 4.5 (A). Then, over the course of 15 years, a strong creative productivity can occur (Figure 4.5 (B)), followed by a period of stability of 20 years (Figure 4.5 (C)), before declining (Figure 4.5 (D)).

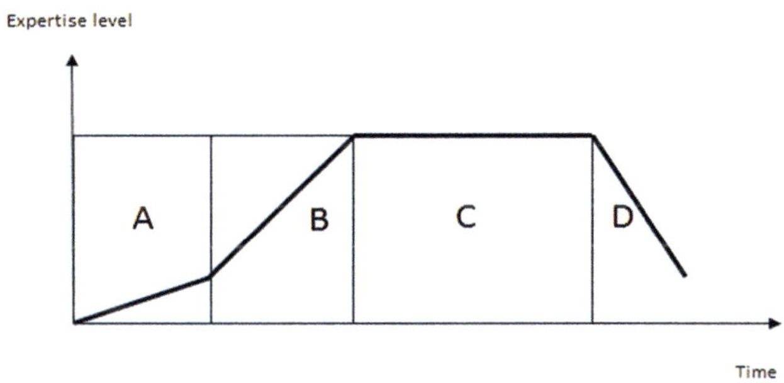

Figure 4.5. *Maximum expertise*

4.2.9. *A production of knowledge that must be interpreted beyond the reasoning at work*

The use of creative rationality is also conditioned by the environment in which designers realize their working activity. Since the 1980s, research carried out on creativity has contributed to the development of a multivariate approach of creativity. According to this approach, creativity involves personal skills (the ability to think in a nonlinear fashion, imagination, motivation, etc.), abilities in a domain (knowledge or expertise, cognitive style, etc.) and environmental factors. In other words, creativity results from factors that are cognitive, conative, emotional and environmental [LUB 03].

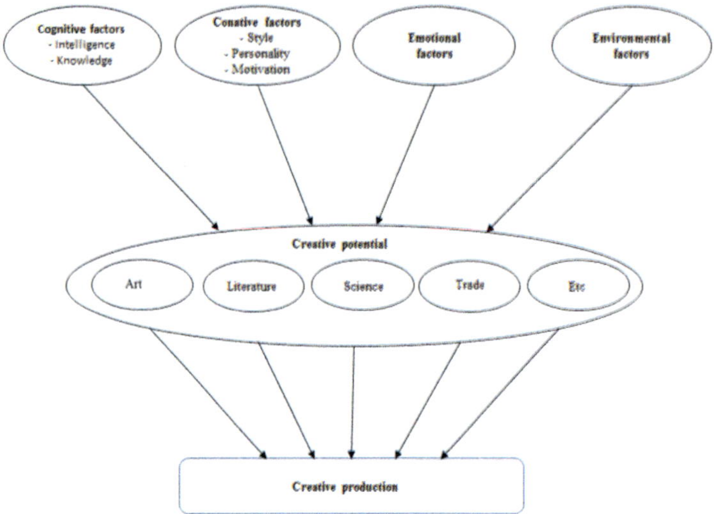

Figure 4.6. *Multivariate approach of creativity*

The multivariate approach of creativity leads us to emphasize, on the one hand, the fact that a high level of creativity can only be reached if most of these factors are present at high levels and, more pertinently to our topic, the relation between environmental factors and creativity. Therefore, realizing that an individual is affected by the numerous influences of his environment, which shape him by changing his feelings and his way of seeing and acting, a certain amount of research has focused on the discussion of the impact of family environment on creativity[25]. Other

25 It seems that family environments organized according to strict rules are not generally favorable to the development of creativity. Undoubtedly, they give to children the representation and experience of an immutable world, where things are such because they must be so. To become creative, a child would need a supportive and relatively uncritical family environment [ROG 54]. Other family factors seem to influence creativity, such as birth order [SIM 75].

works have emphasized the relationship between the type of society (democratic vs. totalitarian) and the development of creativity. According to C. Christensen, J. Dyer, and H. Gregersen, this would explain why Japan, China and several Arab countries struggle *"to challenge the status quo by creating and producing innovations (or winning Nobel prizes)"* [CHR 13, p. 9]. Some have also underlined the consequences of historic events (earthquakes, wars, etc.) or the economic situation (penury vs. wealth, monopolistic situations vs. competitive ones, etc.) on creativity. Finally, others have attempted to highlight the relation between school environment and the development of creativity. We can justifiably hypothesize that a society of creative actors can offer more possibilities of solving the great challenges of today. Therefore, we face the issue of finding out the extent to which the current education system contributes to the deployment of creative rationality.

5

Creative Rationality and the Education System

Considering creative rationality is not only an epistemic but also a practical issue. Conceiving the design process as the driving force of the innovation process entails the theoretical corollary that each improvement of our understanding of the design process will result in a subtler and more pertinent understanding of the innovation process. Its practical corollary is:

– on the one hand, that rationalizing the design process will result in an increased mastery of the innovation process[1];

– on the other hand, that in order to create value, we still need to be able to use creative rationality. In this chapter, we will focus more specifically on the issue of finding out to what extent the education system favors the deployment of creative rationality.

1 It is in this sense that approaches such as functional analysis or design thinking – presented in Chapter 3 – have been adopted.

5.1. Teaching innovation: a political project

As we pointed out in the first chapter, the definition of a policy aimed at favoring and encouraging innovation, since it conditions the dynamism and growth of the economy, has been one of the top priorities of political decision makers for several years[2]. The institutional acknowledgment of the role and challenge of a better understanding and management of innovation resulted, as we have seen, in an explosion of support schemes for innovation. It also resulted in the wish of the Ministry of Higher Education and Research to "*implement mandatory courses focused on innovation and entrepreneurship in every higher-education course*" [MES 12]. As the culture of a population is largely shaped by the education system, mobilizing national and higher education seemed to play a key role.

The desire to educate on innovation follows on from the European Year of Innovation and Creativity in 2009 which, in keeping with the aim of the European Union of becoming a society based on knowledge, led the council and government representatives of the member states to:

– claim that: "*education system must ensure both the development of knowledge and specific skills and that of general abilities linked to creativity such as curiosity, intuition, critical and lateral thinking, problem-solving, testing, risk taking and the ability to learn a lesson from failure, imagination and hypothetical reason, as well as entrepreneurship*" [OFF 08];

2 The correlation between innovation and export performances is a proven fact. Innovating companies export more than companies that do not export. They export to more countries. Their exports grow more rapidly and they are less sensitive to the economic context [BEY 13, p. 33].

– favor the promotion of creativity and the ability to innovate as one of the objectives of the current and future European cooperation in the field of education[3].

There is no doubt that the desire to educate on innovation and creativity is related to the companies' awareness of the significance of innovation culture.

In the last few years, the annual studies *Global Innovation 1000* of the strategy consulting firm Booz & Co, which analyze the R&D spending of the first 1000 global companies, have confirmed that research excellence was not the only factor that affected performances in terms of innovation. The study was carried out in 2011 on nearly 600 senior managers and R&D professionals from more than 400 companies around the globe. The companies had different sizes and belonged to different industrial sectors. The survey revealed that 47% of them regarded innovation culture as a key factor. The conclusion of this study is confirmed by the report "Coping with crisis, the European way" [ERN 13], which underlines that the development of an innovation culture seems to be the second key factor implemented in Europe.

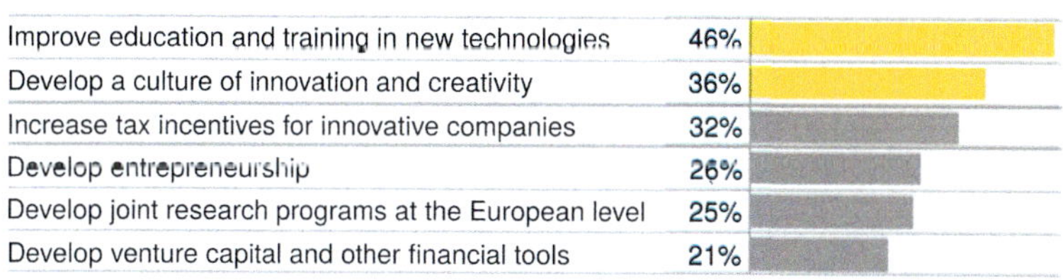

Figure 5.1. *What are the main areas of reform to make Europe a leader in innovation (source: [ERN 13])*

3 It is in this context that, in 2009, France started initiatives in the fields of education and schooling that would contribute to the development of creativity and the ability to innovate.

The conclusion of the report seems corroborated by the survey "Seeking creative candidates: hiring for the future", which was conducted in 2014[4]. This survey revealed that 78% of hiring managers regarded creativity as necessary for economic growth but only 51% of them thought that companies understood the importance of creativity.

As educating on innovation has become a political project, several initiatives have been launched. In 2009, France launched an initiative that would contribute to the development of creativity and innovation abilities, aimed at the education and schooling fields. In 2013, the Prime Minister, eager to make France a land of innovation, made the development of innovation culture one of the four axes that would structure French innovation policies.

5.2. A harmful confusion between innovation and entrepreneurship

It would be very hard not to agree with the project of developing an innovation culture. However, the implementation of this project in the context of the plan "A new order for innovation", launched on the 5th November 2013 in France, suffered from a confusion between innovation and entrepreneurship deriving from the Beylat–Tambourin report [BEY 13] on which it was based.

5.2.1. *The Beylat–Tambourin report*

In said report, the authors pointed out that innovation measures have an entrepreneurial aspect. The authors spelled out that:

4 Only survey with 1068 US hiring managers in July/August 2014.

> *"Each innovation has an entrepreneurial dimension, even in an existing business. Innovation is not a natural process for a human organization. It has to do with the will and determination of one or more individuals. It implies a visionary mind, risk taking, a very developed ability to take the initiative, project culture, and commitment. It requires being comfortable with uncertainties and ambiguities, the ability to identify opportunities that others will never see and to focus on them, to be resolute, persevering, brave, as well as open to ideas and advice"*. [BEY 13, p. 40]

Defined as such, the "entrepreneurial" dimension of innovation closely resembles the qualities necessary for the deployment of creative rationality. However, we can already wonder why we have been led to refer to "entrepreneurial dimensions" instead of a "creative dimension". This observation might seem trivial, unless we put it into perspective with its recommendations.

The authors of the report spelled out 19 recommendations bound to "optimize" public innovation policies in France, which were structured on the following four main points:

– developing innovation culture and entrepreneurship;

– increasing the economic impact of research by transfer;

– contributing to the growth of innovative companies;

– implementing the tools of an innovation public policy.

In relation to the first point, which is the most relevant for our discussion, two recommendations explicitly referred to education.

The first recommendation proposed *"the revision of the pedagogical methods of primary and secondary teaching to develop innovative initiatives*". The second recommendation proposed "*the implementation of a wide-scope program to teach entrepreneurship in higher education*".

However, if we read the Beylat–Tambourin report more carefully, we would realize that the ultimate goal of these recommendations was more "*to favor the students' knowledge of the economic world and companies*" [BEY 13, p. 50] and to develop knowledge about business management/creation rather than learning to use one's creative rationality. In other words, the ultimate goal seems "*the concretization of entrepreneurship*" [BEY 13, p. 50].

This is shown by the fact that if "*teaching basic knowledge (finance, organization, strategy, business method) necessary for the understanding and analysis of entrepreneurial situations*"[5] was mentioned, the notion of creativity was nowhere to be seen, and the reference to innovation even disappeared altogether from the second recommendation[6]. In other words, the creative dimension was eclipsed by the logic of entrepreneurship, as if it were enough to have skills in business management to create so-called innovative companies.

5.2.2. Confusion deriving from J. Schumpeter

The confusion between innovation and entrepreneurship is not new and is rooted in the confusion between entrepreneur and innovator that derives from J. Schumpeter [NOA 13].

---

5 This can also be seen in the shift in meaning from the term "entrepreneurial dimension" to "entrepreneurial situation".

6 We can only find an indirect reference in the discussion of the issue of financial devices aimed at enhancing the creation of innovative companies based on a "start-uppers fund" [BEY 13, p. 53]. It is as if innovation culture seemed equal to entrepreneurship culture, once again raising certain issues.

In his 1912 work, *The Theory of Economy Development* [SCH 12], J. Schumpeter criticized the marginalist way of conceiving economic development. According to the marginalist theory, such a development is based on:

– the figure of the *homo oeconomicus*, who adapts rationally to changes in the environment;

– the way of conceiving the economic cycle on the basis of the principle of equilibrium and adaptation.

He underlined that the static analysis of the marginalist theory is fiction, a utopian vision that never existed since capitalism is never stationary. Not satisfied with this explanation, J. Schumpeter put forward a theory of evolution where innovation was at the source of the dynamics of change in the capitalist economy. In order for these innovations to take place, some economic actors must originate them. However, this agent of change does not exist in the static economic theory. The aim of the economic theory of evolution was to identify it, and J. Schumpeter argued that it should be conceived as this exceptional economic agent he named "entrepreneur".

Let us underline that, according to J. Schumpeter, an entrepreneur does not belong to any definite social group[7]. Similarly, an entrepreneur is not embodied by a physical person. An economic agent may be an entrepreneur one day and then become a manager by adopting a routine behavior. In the same way, in line with what J. Schumpeter said, an entrepreneur cannot be identified with the "simple" figure of a company's creator or director. He is the individual who contributes an innovation, which may temporarily earn him an exceptional income. Paradoxically, some readings

7 Let us underline that, if J. Schumpeter thought that an entrepreneur does not belong to a specific *social* class, he still claimed that there is a "class of entrepreneurs" whose behavior corresponds to a certain number of typical traits [BOU 12].

(undoubtedly a little cursory?) of J. Schumpeter present him as the actual father of the field of entrepreneurship[8].

According to P. Noailles-Siméon [NOA 13], it is precisely when trying to distinguish between the entrepreneur-innovator from the entrepreneur-director that J. Schumpeter confused two terms, a mistake that still affects us today. As Noailles-Siméon underlines, we must distinguish between an innovator and an entrepreneur, as the nature of innovation differs from the nature of common entrepreneurial actions. This is also the theory upheld by C. Christensen, J. Dyer, and H. Gregersen in their work [CHR 13].

5.2.3. *The skills of innovators versus the skills of entrepreneurs*

Based on a comparison between around 500 innovators and 5000 business executives/company directors, C. Christensen, J. Dyer, and H. Gregersen found out five skills that "distinguish actual innovators from traditional executives/directors" [CHR 13, p. 9].

These five skills make up what they named "the innovator's DNA" or "the code for generating innovative business ideas". According to them, the ability to generate innovative ideas is not only a function of the mind (associative thinking)[9] but it also depends on the following behavioral skills:

– Questioning. Innovators never stop asking questions that challenge the status quo.

– Observing. They observe the world around them.

8 An entrepreneurship which is often associated with the creation of businesses in the economic field.

9 What we have called creative rationality.

– Networking. The network of actors represents for innovators a way of finding new ideas.

– Experimenting. Innovators constantly carry out new experiments.

By contrast, as the authors underline, if most managers do not think differently, it is because their four key skills are: analyzing, planning, detail-oriented implementing and self-disciplined executing strictness. In other words, according to C. Christensen, J. Dyer, and H. Gregersen, managers excel at what they call "delivery or executive skills". Their study showed that "great innovators are above the 88th percentile in terms of discovery skills and only at the 56th percentile in terms of executive skills" [CHR 13, p. 20]. This is why the founder of eBay, P. Omidyar, invited J. Skoll, who held an MBA from Stanford, to join him, as the latter had the analytic skills that the former lacked[10].

Thus, the study conducted by C. Christensen, J. Dyer, and H. Gregersen confirmed the idea that the innovators' skills differ from those of entrepreneurs[11].

As these skills are distinct, confusing entrepreneurship culture and innovation culture seems dangerous. Similarly, combining entrepreneurship education with innovation

10 The authors go slightly further, underlining that: "*great companies are not known for their disruptive innovations, as their top management is dominated by individuals chosen for their ability to provide results and not because of their ability to create*" [CHR 13, p. 22].

11 Even if they point out that executive skills are required to introduce the innovative concept to the market. The authors also take care to point out that some individuals, even if few in number, may excel in both types of skills.

education in the same curriculum is risky[12]. If avoiding this confusion seems a necessary condition for the project of educating on innovation, it is not sufficient. We should also ensure that our education system contributes to the deployment of creative rationality.

5.3. School environment and creative rationality

Society and, more specifically, the education system play an essential role in the development of creativity. Creativity in education aims to encourage flexibility, open-mindedness to novelty, the ability to adapt or see that there are different ways of doing something, as well as the courage to face the unexpected [CRO 01].

As G. Berger[13] underlined, three years after the creation of the *Institut National des Sciences Appliquées de Lyon* (National Institute of Applied Sciences Lyon), education has a doubly difficult task: "It should provide techniques and teach students to make efforts to apply them. However, we should never stifle or restrain their enthusiasm. Early childhood often has a fresh imagination, an indefatigable curiosity, a sort of poetic genius that adult scientists or artists struggle to find again. Everything starts with poetry, nothing can be done without technique. But poetry must be so present everywhere that learning mechanisms does not exhaust the

---

12 In *Introduction à l'entrepreneuriat* (2005), A. Fayolle identified three general issues that occur in the field of entrepreneurship: entrepreneurship as an economic and social phenomenon, entrepreneurship as a research field and entrepreneurship as a teaching field. A. Fayolle pointed out that the last issue focuses more precisely on specific knowledge used for entrepreneurship (business plan, starting activities, management, development strategy of young companies, etc.)

13 Both an industrialist and a philosopher specialized in Husserlian phenomenology, he managed higher education at the Ministry of National Education from 1953 to 1960 and was one of the four founders of the INSA in Lyon [CHO 11].

fountainhead of creation" [BER 60]. However, it seems that our education system is struggling to take on this dual task.

5.3.1. *Challenging traditional school*

This remark has been made before. We may hypothesize that it constitutes the reason why, since the end of the 19th Century, traditional school has been put in question. Traditional education became the actual scapegoat of the new education, as we can see in what A. Ferriere, the cofounder in 1921 of the *Ligue internationale pour l'éducation nouvelle*, said:

> "*We created school following the devil's advice.*
>
> *Children love nature: we confined them in closed rooms.*
>
> *Children love seeing that their activity has a purpose: we ensured that it was purposeless.*
>
> *They love moving: we compelled them to keep still.*
>
> *They love handling objects: we introduced them to ideas.*
>
> *They love using their hands: we only made them play with their brain.*
>
> *They love talking: we obliged them to keep silent.*
>
> *They would love to think: we made them learn by rote memorization.*
>
> *They would love to look for science: we served it to them ready-made.*
>
> *They would love to become excited: we came up with punishments (...). Then children learnt what they would have never learnt in other circumstances. They learnt to hide, they learnt to cheat, they learnt to lie*" [FOU 00, p. 14].

The target is the classic education given to learners, in public and private schools, from primary school to university which, according to A. Lumière, "*entails the momentous consequence of annihilating the innate qualities of reasoning, initiative, and observation that can be spontaneously seen in children, when their brain still has not been molded by dogmatic teachings that destroy the spirit of inquiry*" [LUM 35]. A. Lumière backed up his remarks by pointing out that if several great medical discoveries were not made by scientists that were highly skilled in the field[14], it was because their education taught them to keep faithful to theories presented as eternal truths. Similarly, he wrote: "*why are most industrial designs (...) not usually created by former students of the* grandes Ecoles? *They should be in a better position than their less-educated fellow citizens and yet they are not usually innovators, lucky again when, heading companies that they have not created, they prevent them from collapsing, due to the lack of the qualities required to manage them*" [LUM 35, p. 12][15].

The criticism aimed at traditional education gave rise to original pedagogical initiatives. We refer specifically to the pedagogy of Montessori (in Italy) and Freinet (in France). However, if these alternative types of pedagogy were popular in the 1960s, after May 1968 France experienced a "*return to the status quo*", as R. Haby, the Minister of National Education between 1972 and 1978, explained that pedagogical innovation had gone too far. Moreover, the economic crisis at the beginning of the 1970s, together with the increased number of graduates in a society that sees a

14 He reconsidered here the paradox pointed out by C. Nicolle, the winner of the Nobel Prize in Medicine in 1928, in his work *la Biologie de l'Invention* (1932).

15 Let us recall that the observation made by A. Lumière is related both to his career of innovator with his brother L. Lumière and to the fact that he was to innovate the medical field, where he would work with A. Carrel and I. Bérard, without having studied medicine (we owe him especially the invention of tulle gras dressings, for example).

close connection between social and school destiny, only sped up the return to traditional education. Parents, worried about their children in a period of growing unemployment, demanded that their children received the same education they had received themselves. Thus, they campaigned for a return to "old-style teaching". We returned then to a school of discipline, which stifles risk taking and creativity. This is exactly the theory supported by C. Guellerin [GUE 14] in his parable of the "flea trainer".

5.3.2. *The parable of the "flea trainer"*

C. Guellerin's parable of the flea trainer goes as follows:

> *"Once upon a time, there was a flea trainer who had taught his circus animal to jump. He put it in a glass before going on stage and with the encouragement of the audience, he asked the flea to jump. And the flea jumped and jumped, higher and higher, it turned somersaults, it made dangerous and more and more complicated leaps. As the flea learnt to jump better and better, it decided to make new leaps, which even the trainer had not imagined.*
>
> *His role was no longer to constrain, to impose what he could do, but to encourage, to motivate, to see to it that the flea came up with more and more jumps that even the trainer himself could not know, as he was not a flea.*
>
> *Nothing slowed down the imagination of the flea, as it saw the audience become excited about the performance and novelty of the trick staged. Left free to jump over the glass, it dazzled everyone who went to the circus. The time of a trick and the flea 'changed the world' of those who were delighted by such imagination.*

> *However, one day, the flea trainer, while talking to the public and as he wanted to receive some cheering himself, put his hand on the rim of the glass, thus accidentally creating a lid. Hearing the cheering, the flea started to jump. Once, twice, three times, but it banged its head on the hand. It wondered why its trainer was determined to block its movements, but it was its Master. Of course, he knew best.*
>
> *When, after several minutes, the trainer lifted his hand and asked the flea to show all its talent…then the flea jumped and jumped…but how disappointed the public was, the flea could now only jump as far as the rim of the glass. No more dangerous jumps, no more freedom, the Master had just killed the creativity of its Trainee*". [GUE 14]

As we may have understood, the parable of the "flea trainer" leads us to think about the role of teaching and the education system. The current education system is based around conceiving teaching in terms of handing down knowledge, as the learner must stuff himself with knowledge that he will rigorously have to apply if he expects to obtain, at the end of a journey fraught with obstacles, that key to success that is a diploma which, in a country like France, opens up job opportunities[16].

---

16 We should point out here all the ambiguity of the French school system, which campaigns for a school open to all and allowing anyone to reach a good level of education, based on a system bound to bring out the elites of tomorrow, the future captains of industry. As underlined by the report "Towards a more inclusive education system in France": "*If opening education to everyone can be incontestably recognized as an asset of the French education system, step-by-step selection processes during the school career are still in place, they have resurfaced, and occasionally they have even developed further, modeling all of the system, but leaving behind an excessive number of poorly educated young people*" [OEC 15, p. 1].

Thus, this way of conceiving the educational system does not encourage the learner to question knowledge. Teachers seem the holders of immutable knowledge made into absolute truth as they were "educated at the *Ecole normale* (*normale*, namely where they were educated to teach the 'norm' of knowledge)" [TAD 09, p. 43]. Such a viewpoint is all the more surprising as the history of sciences includes examples that show how some theories taken for granted were nonetheless challenged. We can mention, as an example, the shape of the Earth. The idea of Earth as a disc floating on a boundless ocean inherited from Thales was succeeded by the idea of a spherical earth, thanks to Parmenides' works. Similarly, have we forgotten that it was not until Copernicus that the idea that the Earth revolved around the Sun (heliocentrism) replaced the geocentrism that had been predominant since Ptolemy? In a similar fashion, in the medical field, have we forgotten that we owe it to A. Paré for challenging the theory of the four humors elaborated by Polybus in ancient times[17], and that it was not until the beginning of the 18th Century, with W. Harvey's works, that we understood how blood has a circular motion and does not flow only in one direction, as we had been thought since ancient times?

Since past knowledge forms a background that affects our representation of the world, is it so unreasonable to think that in order to find 21st Century solutions to the great challenges of today it is necessary to teach our students to question the knowledge foundations that have been handed down to us and the meaning of objects?

17 Let us recall that, according to Polybus, the diseases plaguing an individual were attributed to an imbalance in the "humors" or liquids flowing in the body: blood, phlegm, yellow bile and black bile. To restore this balance, Hippocratic doctors prescribed to the ill person remedies that aimed to remove the excess humor, especially by bleeding.

Similarly, taking risks without being afraid of failing is the keystone of the creative effort. However, the evaluation culture predominant in most of the Member States of the European Union is contrary to experimentation and risk taking.

In other words, we have inherited an education system that is reluctant to provide our students with the ability to "jump higher", to use the parable of the flea trainer. This remark, naturally, does not apply only to France. This education system is largely perceived as inhibiting creativity: on average 59% of adults in the United States, Great Britain, Germany, Japan and France think this is the case, with no significant difference among these countries[18].

5.3.3. *An education system that kills creativity*

The parable of the "flea trainer" is in line with the remarks of P.M. Perez, affiliated with the Institute of Creativity and Educational Innovation of the University of Valencia. Based on several studies, she revealed that the education system harms the creativity of children and that their ability to create diminishes over the years spent in school. She pointed out that in school we teach children to conform to pre-established patterns, to adopt a convergent rather than a divergent way of thinking. Teachers ask children to provide a precise answer about a determined content that has been taught. And they had better follow the path that has been laid out!

18 However, the French are the least likely to think that being creative may be an asset for society: 53% agree against 76% of Americans, for example. Teachers are also doubtful: France comes second-to-last among the 27 countries of the European Union when people are asked if they think that "the development of the students' creativity plays a significant role in school programs". Only slightly more than 30% believe that this is the case [BEY 13, p. 8].

This idea was hotly debated at the WISE, the World Innovation Summit for Education, which was held from the 4th to the 6th of November 2014 in Doha (Qatar). At the opening conference of the summit, T. Wagner explained that a four-year-old child asks around a hundred questions a day. When he is 7, he starts understanding that it is better to be able to answer questions rather than asking them.

Sir K. Robinson reached the same conclusion [ROB 06]. In his conferences, he explained that the education system is largely based on a way of conceiving intelligence inherited from the enlightenment, according to which being intelligent means being able to reason according to deductive logic and master the humanities. He pointed out that children take risks, improvise, and are not afraid of making mistakes but, with age, they gradually lose their ability to create.

Therefore, the role of school should be urgently reconsidered. It is not enough to only memorize facts and theories. As Montaigne (1533–1592) said, "*Care be taken to choose a guide with a well-made rather than a well-filled head*". Accumulating knowledge is not an end in itself, and we should still be able to make something out of it and consider the meaning of what we do. In this respect, we agree with J. Ferry, when he pointed out that ultimately "*it is not a matter of embracing everything it is possible to know, but to learn well what we are not allowed to ignore*" [FER 82]. From this perspective, learning to create and handing down the ability to create seems a prerequisite to meet the great challenges of the 21st Century. However, we face the issue of finding out how school could be changed.

In France, national education highly encourages innovation. We can find in each academy advisers on research, development and innovation. Similarly, the frame of reference that guides teacher training [FRE 13] strongly emphasizes the teachers' need to develop "pedagogical innovation approaches aimed at improving practices"

[WAT 14]. This phenomenon does not spare universities, whose organization charts are witnessing the growth of innovation and development branches (University of Lyon 3), such as the "research and innovation branch" at the University of Rennes 1, or new functions like the "assistant director of innovation" at the INSA Lyon.

However, as P. Watrelot (2014) underlined in his column *Ecole et Innovation: je t'aime moi non plus*: "*If National Education develops a discourse on innovation and attempts to promote it, we can still question the ability of a bureaucratic and centralized system to produce innovation*". Besides, mentioning E. Moring, the author recalled that: "*We should constantly rely on an active avant-garde. There is never consensus before innovation. We do not make progress based on an average opinion that is mediocratic rather than democratic*" (Morin in [WAT 14]).

Numerous initiatives, like the "getting your hands dirty" project started in 1996 by G. Charpak (the winner of the Nobel Prize in physics in 1992) and the Académie des sciences, have been launched to update the way of teaching sciences and technology in primary schools. Breaking with the "empiricist" teaching of sciences, based on classes during which students must absorb the knowledge handed down by the teacher, "getting your hands dirty" introduces a "constructivist" type of learning where students demonstrate creativity and autonomy to rediscover scientific facts on their own.

E. McWilliam, P. Pronnick, and PG. Taylor [MCW 08] agree. In their article, "Re-designing Science Pedagogy: Reversing the Flight from Science", they "*implore Faculty to redress the issue by engaging in scientific teaching, i.e. in teaching science as it is practice*" [MCW 08, p. 226]. According to the authors, integrating the teaching of creativity is a way of updating scientific curricula and pedagogies, helping to change the direction of modern science for the following four reasons:

– young people are more engaged by active tasks than with a passive consumption approach to transfer of core knowledge;

– it is boredom, not rigor, that disengages them – the difference between static and dynamic sources of knowledge;

– creativity is not the antithesis of scientific rigor, but the core business of scientific thinking;

– we now have new understanding of creative pedagogies that makes teaching strategies visible and effective.

These strategies can build academic, digital and social capacity simultaneously, and this is the new core business of the science educator [MCW 08, p. 228].

It is for similar reasons that in the last few years, we have spoken in favor of and striven towards the rehabilitation of creativity rationality in the training of engineers.

5.4. Rehabilitating creativity rationality in the training of engineers

As T. Gaudin, an expert in forecasting and innovation, underlined, "*an engineer (…) is seen as a factor of production, not as an innovator and disruptor that raises actual problems (pertinence) unceremoniously (impertinence)*", adding that "*rationalizing and creating are complementary, yet opposed, approaches: on one hand, sectioning, a laborious deconstruction into elementary difficulties, and the reference to solutions that have already been tested; on the other hand, the emergence of a form, an artistic composition that evades logic, and a creative step*" [GAU 84, p. 135]. However, we should point something out. In engineering schools, teaching creativity plays a minor role in the education offered. These schools are, in this respect, the heirs of a Western tradition that confines (as we saw in Chapter 2) technique outside the logos and turns it into nothing more than the mere application of a science that is external to it.

This remark has been made before, and H. Simon referred to it at the end of the 1960s in his work *The Sciences of the Artificial*. According to him, engineering schools have become schools that teach physics and mathematics. He stated that the use of the adjective "applied" only hides this fact but cannot change it. This does not mean that designing is taught as such in these schools.

The observations of H. Simon were updated at the beginning of the 1990s in the report "Improving engineering design, designing for competitive advantage" [NAT 91]. In the chapter called "Improving Engineering Education", the authors of the report underlined that designing is an activity that characterizes engineers, but that courses do not cover the basics and the nature of designing.

The significance of the correspondence between the education of engineers and their future activity led the Journal of Engineering Education to reconsider its objectives in 2003 [FEL 05]. It was this pressing consideration that in 2006 led the Special Report of the Journal of Engineering Education called "The research agenda for the new discipline of engineering education" to appeal to the nation: "*our nation needs to make the critical research investments that will transform today's educational system into the pre-eminent paradigm for engineering education and ensure that the U.S. maintains its leadership role in addressing the global challenges of the future*" [JEE 06, p. 261].

The relative concealment of designing in courses is damaging, as engineering schools have elaborated an education model that apparently favors the model of analytic reason to that of creative reason: "(...) *in the history of western thought, 'meaning, respectable meaning, was identified with the logical thinking of humankind, while human imaginative thought was identified with the animistic, the irrational, the illogical, the instinctual, the repressible, and ultimately the dangerous*" ([MUR 86, p. 235] in Policastro and Gardner [POL 05]). Isn't the issue faced by engineering schools precisely to favor this creative rationality and distinguish it from analytic rationality, which often caricatures it or in any case masks it?

As G. Berger had already underscored, "*our relationship with technology is not a mere application of science, as for inventors it is not enough to apply and reproduce*" [BER 58, p. 6]. It is precisely this remark that led Berger[19] to refuse to consider theoretical intelligence as the only actual form of intelligence and to reject any competitive entrance examinations at the INSA Lyon, which opened its doors in 1957. Exams select on the basis of *theoretical intelligence* (which allows students to obtain the best marks in the main university tests), disregarding *imaginative intelligence* and *practical or concrete* intelligence. It is precisely the awareness of these three forms of intelligence that helped establish recruitment processes that can take into consideration both intelligence (in relation to the marks obtained in essays during the year and in the high school diploma) and character (during interviews) in order to assess personality more thoroughly. What made the INSA Lyon stand out in 1957 was not the absence of entrance examinations as much as the recruitment processes it implemented [CHO 11].

19 One of the four basic principles of the *Institut National des Sciences Appliquées* in Lyon.

If, owing to the creation of the INSA Lyon, the significance of imaginative intelligence was pointed out in France at the end of the 1950s, it must be said that the issue of the meaning of what is taught, especially in Social and Human Sciences, is rarely raised. We have already indicated how a way of teaching Social and Human Sciences imported from preparatory classes based on exam-like grading scales and allowing students to think only in very normative terms[20] does not leave much room for the expression of creativity [FAU 13]. Not only does this type of teaching seem unfavorable to "the art of unexpected connections" [BER 58, p. 6] which, as G. Berger underlined, were vital for inventions, but the lack of debate about the meaning of said teaching leads to contradictory teachings given to students, i.e. "learn to respect the rules" but "be innovative!".

5.5. Towards the pedagogy of adventure

As we have seen, the order to innovate involves the rehabilitation of creative rationality in the curriculum of engineers. This rehabilitation represents a significant issue and compels us to reconsider the Western metaphysical tradition that invalidated this type of rationality. Rehabilitating creative rationality in relation to the education curriculum of engineers leads us to consider what we have called the "pedagogy of adventure" [FAU 11], which attempts to consider the deployment of creative rationality. After presenting two keystones, namely the development of observation and otherness, we will outline a possible way of moving from the idea of the pedagogy of adventure to its implementation.

20 It would be useful to consider to which extent it is actually possible to refer to personal reflection when this reflection is dictated by teachers.

5.5.1. *Observing to innovate*

The first keystone of the pedagogy of adventure may come as a surprise to the fans of adventure. It involves being able to observe the world around us, as it is and not as we imagine it.

We mention, as an example, the emergence of the Tata Nano. According to the story, R. Tata, witnessing a relatively common scene in India, i.e. a whole family riding on a scooter, considered the possibility of conceiving the "people's car" of the 21st Century.

Figure 5.2. *A family riding a scooter (source: http://www.innovation-creative.com/index_7.html#tata)*

In 2003, R. Tata announced the manufacturing of the future Tata Nano, a car that would be sold for the target price of 100,000 rupees, namely around 2000 dollars. For four years, 500 engineers worked on a way of manufacturing this car. Each part was studied and reinvented, starting all over again. The model was revealed in 2008 during the ninth motor show in New Delhi.

Developing observation skills aims to discover what the untrained eye cannot see. As M. Proust said, "*The real voyage of discovery consists not in seeking new landscapes, but in having new eyes*".

This approach, in line with design thinking, encourages learners to behave like anthropologists. It first asks our students to walk in and observe an environment that is familiar to them. This untrained observation allows them to become aware of everything they have never seen. It is thus that they discover:

– Uncommon objects: at the INSA's Humanities Center Lyon, there is still a disconnected wired telephone (Figure 5.3).

– Problematic situations: situated in the outskirts of the Lyontech la Doua campus in Villeurbanne (France), the *Boulevard du 11 novembre* has a sidewalk that, following the logic of the creation of a sustainable city, favors a soft mode of transport, namely cycling, but seems to have forgotten that the traditional purpose of a pavement is pedestrian traffic! (Figure 5.4).

Figure 5.3. *A telephone booth disconnected from the network, located at the INSA's Humanities Center in Lyon (source: picture taken by the author)*

Figure 5.4. *A sidewalk that relegates pedestrians to the edges (source: picture taken by the author)*

Such an observation may lead us to identify problematic situations that can provide value-added concepts. We should underline that observation, such as it is, defined and practiced in social sciences also provides key information to the designer, so that he can better delineate his concept.

Observation turns out to be a very effective method to better understand and assess what users do with technology, or even what they do in general[21]. More precisely, observation, whether covert or overt, participating or not, helps us come close to and understand worlds that are foreign to us, get rid of our prejudices, but also appreciate the pertinence of our concept in relation to needs identified as real. Thus, and as T. Brown underlined, "*innovation is powered by a thorough understanding, through direct observation, of what people want and need in their lives and what they like or dislike about the way particular products are made, packaged, marketed, sold and supported*" [BRO 08, p. 86]. For example, it is by observing a consumer who lifts several times his 9-kilo mineral water pack on his

21 Participating observation is especially productive in contexts where words are absent or conventional and do not allow us to go beyond clichés and stereotypes.

way home that the idea of attaching a handle to the pack, to make them easier to pick up and carry, may be born.

Because on a daily basis we are all partially sighted, the pedagogy of adventure encourages us to rediscover the world around us, to track down disappointed expectations and practices so as to make them actual medium of innovation. Thus, the pedagogy of adventure encourages us to free ourselves from the idea that in order to innovate it is enough to listen to what customers have to say. Innovating involves being able to observe users with the aim of providing them artifacts that they would have never imagined they needed. However, the observation approach will be all the more rich, as the observer will strive to analyze and understand the practices and behaviors of the users, whom he will question, or, in other words, the meaning of said practices and behaviors.

Training to observe, consequently, involves not only teaching an engineering student what should be noted, how to collect data, and which are the precautions and bias of the observation method, but also:

– to see daily things with new eyes;

– to be surprised;

– to question why situations they observe are that way.

Focusing on the issue of meaning is vital, as artifact creation is not an end in itself. We do not create objects independently of their context of use, but objects that can create possible actions for a user. For example, it is by questioning why users do not take the path laid out by town planners to go from one building to another, taking instead their own path[22] across the lawn, that we discover the

22 We can find here the idea of poaching put forward by M. De Certeau. Users are not sheep that "are only free to graze on the ration of simulacra that the system allots to everyone" [DEC 90, p. 240].

meaning of this behavior. Users walk across the lawn because they can reach the other building faster. Let us nuance this claim some more: what do we gain by taking the shortest path? This line of questioning leads us to become aware that this behavior is not independent of the values of the society in which we live. The present is marked by sped-up time and increased speed. We can find the premises of these phenomena at the beginning of the 19th Century with the advent of the first steam locomotives. More recently, the revolution of the new information and communication technologies (NICT) led to an exponential extension and acceleration of all forms of exchange. A sped-up time transforms, in Western societies, not only how we live, but also our relationship with the world. This era helps make saving time a unit of value: do we not commonly hear that time is money?

Figure 5.5. *The creation of a side path*

The pedagogy of adventure, by favoring observation in anthropological terms, induces a sort of wandering that makes it possible to free ourselves from ethnocentric tendencies, to get rid of our beliefs and stereotypes, and to disrupt our daily blindness. To speak in other terms and think differently, we should learn to look with new eyes. This pedagogy also allows us to question the way we conceive the world in which we live and create. More precisely, the pedagogy of adventure allows us to see how the use of creative rationality must be conceived beyond the reasoning at work. Consequently, we are led to discard the theorization of design thinking put forward by H. Simon (limited rationality) and A. Hatchuel (C-K theory). Understanding the deployment of creative rationality involves leaving aside approaches that we will call "egocephalocentric" in favor of an approach that takes into consideration the crucial issue of the social framework of the production of knowledge and thus contextualizes this rationality in its historical, cultural and anthropological dimension, namely the field in which it operates [CHO 10].

5.5.2. *Otherness: recognizing the other*

The second keystone of the pedagogy of adventure, which is closely associated with the previous one, proposes to educate on otherness.

Otherness is not an issue of theoretical knowledge. It is first of all a matter of meeting. Spontaneously, several people fear otherness, since meeting the other involves facing novelty and therefore dealing with the unknown. This is why those who fear otherness are often "fearful" and often opposed to innovation. Educating on otherness, consequently, aims to make people aware of what the other brings us. This is precisely what R. Lester and M. Piore showed in their work *Innovation: the missing dimension* [LES 04].

Trying to understand the microeconomic foundations of innovation based on case studies focusing on companies that belong to different lines of business (cell phones, blue jeans, cars and medical equipment in this case), R. Lester and M. Piore pointed out that innovation results from two processes that are simultaneously complementary and largely contradictory, namely:

– an analytic process that involves problem-solving;

– an interpretative process that involves creativity.

The authors showed that the interpretative process creates interpretation opportunities that can bring about certain possibilities: "cell phones emerged out of a conversation between members of the radio and telephone industries; medical devices emerged out of a conversation between academic scientists and medical practitioners with clinical experience [...] In each case, the manager's role was to remove organizational barriers that would have prevented these conversations from taking place" [LES 04, p. 52]. The authors wondered then how this interpretative process could be improved. Starting with the metaphor of a cocktail party, the authors pointed out that the individuals involved must come from different backgrounds: *"But just as a cocktail party would be tedious if all the guests had the same background and agreed about every idea, a business organization from which ambiguity has been removed is unlikely to produce anything very innovative and interesting. (...) Everyone who has ever attended a real cocktail party knows that what makes these events interesting initially is that they bring together people from different walks of life"* [LES 04, p. 69].

The importance that the authors attached to diversity[23] and meeting in the innovative phenomenon is shared by many. C. Villani, in his work *Théorème du vivant*, reached the same conclusion and described the way in which *"subjects mix during conversations, among researchers pursuing different mathematical goals around a coffee*

23 We can find here the results of the abundant literature underlining the role of diversity in innovation: *"diverse teams of people are more creative than homogeneous teams"* [LAN 08].

machine or in the corridors, without fearing thematic barriers" [VIL 13, p. 24] and how the individuals he met during his stay at Princeton in 2009 led him to stray from his scientific trajectory in a way that he could have never imagined.

Several examples go even further and reveal the role of unexpected meetings in the emergence of innovations. An unexpected meeting is precisely what gave birth to a drug against kidney transplant rejection. The marketing of this drug depended on an unexpected encounter of two researchers in a "*rendezvous so boring that allowed the two researchers to 'meet' on a scientific level (...), a lucky encounter, from the point of view of its conditions, between two scientists. We witness here the beginning of a peculiar story, which involves the creation of an unprecedented drug that uses a monoclonal antibody no longer used for diagnostic purposes, as has been the case for years, but with a therapeutic aim*" [BIB 91, pp. 277–278].

We can also mention the example of the Biostyr water treatment method invented by the CGE (Compagnie Générale des Eaux) following the development of the European law on the elimination of the nitrogen and phosphorus contained in the wastewater discharged into rivers by wastewater treatment plants in 1985. The story of this invention reveals that the research center Anjou Recherche, located in Maisons-Laffitte, played the role of "*breeding ground for the cross-fertilization of basic research and practical testing, with subsidiaries made complementary by the interdependence of their specializations. It plays the role of forum for discussion, encouraging the creativity of the members and leadership thanks to multiple un-hierarchized interactions*" [ASS 02, p. 26].

These examples allow us to confirm the theory upheld by G. Horowitt and V. Hwang in their book *The Rainforest: The Secret to Building the Next Silicon Valley* [HOR 12]. The authors underlined that Silicon Valley works like the Amazon forest and that innovation is like a weed. It grows in environments that do not look like farmlands but in lush tropical forests that are rich in exchanges.

According to us, considering the role of otherness compels us to educate on creativity in a different way. Techniques such as brainstorming aim only at multiplying ideas. However, we think that educating on creativity involves first becoming aware of the other. By meeting the other, students will be able to face other worldviews, paradigms and cultures that will necessarily challenge their cultural certainties and boost their creativity.

In return, teachers must invent a pedagogy that is itself adventurous and considers knowledge not as a process of duplication but as an element handed down and compelling students to challenge their certainties and transform them. Thus, knowledge transfer becomes also a knowledge adventure.

5.5.3. *How to move from the idea of pedagogy of adventure to its implementation?*

As discussed above, trying to make creative rationality an issue of knowledge and practice involves the implementation of an adventurous pedagogy that encourages students to question and transform their certainties. Thus, knowledge transfer is coupled with a knowledge adventure.

Such a pedagogy encourages the invention of new forms of transmission. It may lead us to leave familiar places like the classroom to settle temporarily in different places such as:

– Technical museums, like the Printing Museum or the Institut Lumière Museum in Lyon:

a)

b)

Figure 5.6. *a) Printing museum in Lyon b) Institut Lumière in Lyon*

Visiting these museums allows us to discover the history of the creation of J. Gutenberg's typographic printing press and the brothers A. and L. Lumière's invention of the cinematograph. These visits help to demonstrate the mechanism of knowledge crossover at work in their innovation.

– The Institute of Contemporary Art in Villeurbanne:

These places reveal several forms of creativity. The transfer of knowledge may thus take place in a new way by confronting students with concrete works.

As the pedagogy of adventure aims to allow students to come across different worlds, some meetings are organized with a designer, B. Buffalino, who explains to them his creative process based on his works.

a) b)

Figure 5.7. *a) Urban evasion, the aquarium telephone booth, Lyon, 2007 (source: www.benedettobufalino.com). b) Urban foosball table, Lyon, 2014 (source: www.benedettobufalino.com)*

In the last three years, we decided with my colleague M. Faucheux to go even further by introducing our students to an even more distant universe, which some may call exotic, i.e. the world of fashion[24]. To this end, we decided to get L. Guillot involved in our approach. L. Guillot is a custom wedding gown designer in Lyon who has been working in the field of beauty queens since 2005. In the first year, her involvement consisted of presenting her creative process for the Official Regional Outfit "Miss Rhône-Alpes" for the election of Miss France 2015. In concrete terms, she explained to students how, based on the specifications of the

24 To understand properly the "cultural shock" that such an approach may cause, let us point out that this pedagogical tool was used on a mostly male audience (80%).

Miss France society[25], she explored the imagination associated with the Rhône-Alpes region, chose to highlight a major innovation of the region, namely the Lumière brothers' cinematograph, and represented this choice in the creation of her dress (see Figure 5.8).

Figure 5.8. *Sketch of the regional outfit of Miss Rhône-Alpes (source: L. Guillot creation)*

25 These specifications stated that the regional costumes created should be entirely revisited and more modern, representing the regional identity and drawing their inspiration either from folklore or from the specialties, history and cultural heritage of each region.

Meeting L. Guillot allowed our students to question the links between technical and artistic creativity. This meeting became then a pretext to tackle with them the associations between imaginary and innovation. As M. Chouteau and C. Nguyen [CHO 15] underlined, the imaginary is a reservoir of ideas, desires, utopias and inspiration favorable to innovation. The imaginary, thanks to intentions, worldviews, etc., conveys a meaning that, as we have seen, is necessary for innovation. Considering that technical objects represent how the imagination becomes concrete encourages us besides to compare the designer's imagination with the user's, to the extent that we do not sell merely an object, but also the culture and the imaginary that surround it: "*smartphones are not used only for communicating, they are also bought because they allow us to maintain strong family ties thanks to different apps (Skype, SMS, MMS, etc.)*" [CHO 15].

The following year, we decided to ask our students to create on their own a dress that symbolized the Rhône-Alpes region and the world of engineers. By being compelled to leave their field of knowledge (and comfort zone), our students experienced the adventurous transgression[26]. This exercise encouraged them first to work with the imagination associated with the Rhône-Alpes region and the engineering world. These types of imagination were then assembled into an ambiance mood board (Figure 5.9).

26 We should point out that, as was the case for the project "Surprise yourselves, surprise us", which the students led as a group, even the teacher went on an adventure as he did not know where they would end up and there was no right or wrong, no boundary, end, or self-evident and predictable fact. In short, the students surprised us!

Figure 5.9. *Ambiance mood board (work created by M. Mamadou Dagra, G. Gentile, A. Margaillan, and C. Mendez Antúnez, Lyon INSA, February 2017). For a color version of the figure, see www.iste.co.uk/forest/innovation.zip*

The previous ambiance mood board inspired the creation of the dress model "The gift of the Rhone" (Figure 5.10).

The sketch of the dress was based on the idea of the Rhône as "cradle" of the cultural and industrial development that took place all along the Rhône valley. The outer part of the dress recalled the natural elements of the region, from mountains to valleys, where the Rhône, represented by the ribbon, brought everything together. Second, the inner part of the dress revealed the technological progress and the richness of the region, which were made possible by the resources of the Rhône. Thus, we can see a cinema film roll (which refers to the Lumière brothers' invention of the cinematograph), silk balls (which refer to the history of Lyon silk-workers) and a movable type (which refers to the history of printing in Lyon). The transition between natural and artificial was highlighted by the large gears on the bust. Behind, the movable types RA referred to the initials of the region. We should point out that the strap marked the contrast between natural and artificial, associating the presence of an edelweiss with that of the gears.

Figure 5.10. *Sketch of the dress "The Gift of the Rhone" (work by M. Mamadou Dagra, G. Gentile, A. Margaillan and C. Mendez Antúnez, Lyon INSA, February 2017)*

Finally, the creation of the dress was a pretext to turn creative rationality into a practical issue. With this exercise, students were urged to leave the comfort zone of their field of knowledge to explore unfamiliar worlds. They realized that they could be creative as long as they allowed themselves to be creative. They became aware that what is important is not "thinking out of the box", as in reality the box is nothing more than a mental construct. The dress exercise compelled them to leave their inhibitions behind: they had to accept the fact that they had to create a dress as a group and present their approach and work to the rest of the group and L. Guillot.

We can see from this that the pedagogy of adventure seems a way of giving free rein to the learner, favoring the indiscipline of students[27]. Consequently, it breaks with the way of conceiving school as the place where we must abide by the rules and expectations in favor of a representation of school as a breeding-ground for adventure. It is a type of pedagogy that allows students to structure themselves in a process of "co-naissance" (co-birth) that educates and changes them along the way.

Such a type of pedagogy leads us to reconsider our evaluation system. This system favors analytic reason to ingenious reason, which belongs nowhere in an evaluation framework where everything is known *ex ante*. What distinguishing problem-solving from designing is precisely the fact that, in the first case, the set of possible solutions is completely specified. As the solutions are already there, it is enough to go look for them. Analytic reason is mobilized to solve problems whose data constitutes the narrow

27 Let us make no mistake about the meaning of this remark. By the indiscipline of the students, we mean students who dare to leave the comfort zone of their discipline and cross knowledge domains. This does not mean denying knowledge. As M. Serres pointed out: "*every institution, every exercise, and every type of education requires a structure. However, every invention requires us to discard this structure*" [SER 15].

framework within which it moves. It is not involved in a process that leads it to exert a creativity that can go beyond the problem, contextualize it, and tackle it on a meta-level that relativizes it, helping us see it from a different perspective and imagine different types of logic and solutions radically.

The pedagogy of adventure therefore involves the indiscipline of students and teachings. Such a pedagogy compels us to devise human and social sciences that are not imported as such from universities, but become meaningful in engineering schools. Thus, they necessarily lead us to cross different knowledge domains at the cost of an effective interdisciplinarity and draw a new map of knowledge, where knowledge is an archipelago that resembles the splintered geography of the Odyssey.

Conclusion

In this book, we have attempted to highlight the issue of considering innovation in different terms. If we agree to discard the linear and hierarchical model of innovation and to consider innovation through the prism of Artificialism, we are led to identify new possible means of action and update a form of thought that has long been disqualified by history, i.e. creative rationality.

As we have pointed out, creative rationality, which is a form of thought at the very core of innovation, is a relational type of thinking which works according to the principle of what we have named adventurous transgression. Starting from there, how can we leave aside the injunction to innovate in favor of an effective ability to innovate?

We support the theory according to which it is vital that our education system does not kill the learners' creativity. There are naturally no good solutions, but plenty of bad ones. Obviously, structuring teaching into watertight and hierarchized disciplines stifles the deployment of creative rationality, and so does importing preparatory class teachings for top-ranking higher establishments as they are into engineering schools. The lesser of two evils is to offer possibilities of meeting and to favor cross-fertilization. Such a solution promotes the adventure and the virtues of otherness and of the unexpected.

Finally, let us underline that making creative rationality an educational and practical issue also helps us discard a purely functional way of conceiving technology. The use of creative rationality highlights the associations between technology and society. Education on creative rationality is then part of the development of a technical culture [CHO 15].

Bibliography

[AAL 88] AALTO A., *Alvar Aalto: de l'œuvre aux écrits*, Editions du centre Geoges Pompidou, Paris, 1988.

[AGH 04] AGHION P., COHEN E., "Education et croissance", Report for the Conseil d'analyse économique, Paris, 2004.

[ALB 05] ALBERT R.-S., RUNCO M.-A., "The concept of creativity: prospects and paradigms", in STERNBERG R.J. (ed.), *Handbook of Creativity*, Cambridge University Press, Cambridge, 2005.

[ALT 00] ALTER N., *L'innovation ordinaire*, Presses Universitaires de France, Paris, 2000.

[ARC 65] ARCHER L.-B., *Systematic Methods for Designers*, The Design Council, London, 1965.

[ARC 84] ARCHER L.-B., "Systematic method for designers", in CROSS N. (ed.), *Development in Design Methodology*, John Wiley & Sons, New York, 1984.

[ASS 02] ASSENS C., BARONCELLI A., FROEHLICHER T., "Processus d'émergence et de diffusion d'une innovation technologique au sein d'un réseau de firmes", *Cahiers de recherche du GREFIGE*, vol. 2002–01, 2002.

[AUG 03] AUGER P., "Favoriser la créativité individuelle pour développer l'innovation", *Actes du congrès AGRH,* availabe at: https://www.agrh.fr/assets/actes/2003auger007.pdf, pp. 145–166, 2003.

[BEC 92] BECHTEL G., *Gutenberg et l'invention de l'imprimerie,* Fayard, Paris, 1992.

[BER 58] BERGER G., "Humanisme et technique", *Revue de l'enseignement supérieur*, no. 1, pp. 5–10, 1958.

[BER 60] BERGER G., Discours L'université dans la cité, Perpignan, 29 juin 1960.

[BER 15] BERDIER C., "La gestion des déchets ménagers dans l'habitat social vertical à l'épreuve de l'innovation technologique", in FOREST J., HAMDOUCH A. (eds), *Quand l'innovation fait la ville durable*, Presses Polytechniques et Universitaires Romandes, Lausanne, 2015.

[BEY 13] BEYLAT J.L, TAMBOURIN P., L'innovation un enjeu majeur pour la France: dynamiser la croissance des entreprises innovantes, Report on behalf of the Ministères de l'Enseignement Supérieur, du Redressement Productif, Paris, 2013.

[BIB 91] BIBARD L., "Histoire d'une innovation: le cas d'un médicament", *Revue économique*, vol. 42, no. 2, pp. 273–300, 1991.

[BIG 29] BIGELOW J., *Elements of Technology*, Gray & Co., Boston, 1829.

[BIR 12] BIRRAUX C., LE DEAUT J.-Y., L'innovation à l'épreuve des peurs et des risques, Report, l'Office Parlementaire d'Evaluation des Choix Scientifiques et technologiques, 24 January 2012.

[BLE 94] BLESSING L., *A Process-Based Approach to Computer-Supported Engineering Design*, Black Bear Press Ltd, Cambridge, 1994.

[BON 02] BONNOT T., *La vie des objets. D'ustensiles banals à objets de collection*, MSH, Paris, 2002.

[BOO 11] BOOZ & COMPANY, Why Culture Is Key: The 2011 Global Innovation 1000 available at: http://www.strategyand.pwc.com/media/file/Strategyand-Global-Innovation-1000-2011-Webinar.pdf, 2011.

[BOS 08] BOSCHMA R.A., Constructing regional advantage: related variety and regional innovation policy, Report for the Dutch Scientific Council Government Policy, 2008.

[BOU 12a] BOUTINET J.P., *Anthropologie du projet*, Quadrige, Paris, 2012.

[BOU 12b] BOUTILLIER S., UZUNIDIS D., "Schumpeter, Marx et Walras. Entrepreneur et devenir du capitalisme", *Revue Interventions économiques*, vol. 46, available at: http://interventionseconomiques.revues.org/1690, 2012.

[BRO 94] BROOKS H., "The relationship between science and technology", *Research Policy*, vol. 23, pp. 477–486, 1994.

[BRO 08] BROWN T., "Design thinking", *Harvard Business Review*, June, pp. 84–95, 2008.

[BUR 14] BURKUS D., *The Myths of Creativity*, Jossey Bass, San Francisco, 2014.

[CAR 97] CARRIER C., *De la créativité à l'intrapreneuriat*, Presses Universitaires du Québec, Québec, 1997.

[CHO 10] CHOULIER D., COATEANA E., FOREST J., "The engineering design CK theory: contributions and limits", *22nd International Conference on Design Theory and Methodology*, Montréal, 2010.

[CHO 11] CHOUTEAU M., FOREST J., NGUYEN C., "Il était une fois l'INSA... Un réseau d'hommes réformateurs", in D'ENFERT R., FONTENEAU V. (eds) *Espaces de l'enseignement scientifique et technique. Acteurs, savoirs, institutions, XVIIe-XXe siècles*, Hermann, Paris, 2011.

[CHO 15] CHOUTEAU M., NGUYEN C., "L'imaginaire dans la formation au métier d'ingénieur. Quelques exemples d'enseignements à l'INSA de Lyon", Colloque Création-créativité et innovation dans la formation et l'activité de l'ingénieur, CNAM, 3 December 2015.

[CHO 17] CHOUTEAU M., FOREST J., NGUYEN C., "Quand la culture d'innovation fait écran à la culture technique", *Technologie et innovation*, vol. 17-4, available at: https://www.openscience.fr/Quand-la-culture-d-innovation-fait-ecran-a-la-culture-technique, 2017.

[CHR 13] CHRISTENSEN C., DYER J., GREGERSEN H., *Le gène de l'innovateur: Cinq compétences qui font la différence*, Pearson France, Montreuil sous Bois, 2013.

[CIA 07] CIADT, Les pôles de compétitivité, available at: http://competitivite.gouv.fr, 2007.

[CIS 12] CIS, Community Innovation Survey 2008–2010, Central Statistics Office, April, 2012.

[COM 09] COMMISSION OF THE EUROPEAN COMMUNITIES, Design as a driver of user-centred innovation, SEC(2009)501, 2009.

[COM 10] COMMISSIONN EUROPEENNE, Initiative phare Europe 2020: Une Union de l'innovation, Communication de la commission au parlement européen, au conseil, au comité économique et social, SEC(2010) 1161, available at: http://eur-lex.europa.eu/legal-content/FR/TXT/?uri=celex:52010DC0546, 2010.

[CRO 01] CROPLEY A.J., *Creativity in Education and Learning, A Guide for Teacher and Educator*, Routledge, London, 2001.

[DEC 90] DE CERTEAU M., *L'invention du quotidien*, Folio essai, Paris, 1990.

[DEL 14] DE LA PORTE X., "L'innovation et la croyance un peu magique qu'elle va nous sortir de la crise", *Rue 89*, available at: http://tempsreel.nouvelobs.com/rue89/rue89-ce-qui-nous-arrive-sur-la-toile/20140109.RUE1254/l-innovation-et-la-croyance-un-peu-magique-qu-elle-va-nous-sortir-de-la-crise.html, 2014.

[DEL 15] DELENEUVILLE M., "Le vrac, un argument de choix pour le commerce alimentaire", *Journal Du Net*, available at: http://www.journaldunet.com/economie/distribution/1167306-le-vrac/, 2015.

[DER 90] DERTOUZOS M., LESTER R., SOLOW R., *Made in America*, InterEditions, Paris, 1990.

[DES 15] DESIGN COUNCIL, Innovation by Design How Design Enables Science and Technology Research to Achieve Greater Impact, Report, available at: http://www.designcouncil.org.uk/sites/default/files/asset/document/innovation-by-design.pdf, 2015.

[DET 74] DETIENNE M., VERNANT J.-P., *Les ruses de l'intelligence, la mètis des Grecs*, Flammarion, Paris, 1974.

[DOR 15] DORTIER J.-F., "Bouillons de culture", *Les grands dossiers de Sciences Humaines, Special Edition, Innovation et créativité*, no. 38, pp. 11–13, 2015.

[ERN 13] ERNST & YOUNG, Coping with the crisis, the European way, available at: http://www.ey.com/Publication/vwLUAssets/European-Attractiveness-Survey-2013/$FILE/European-Attractiveness-Survey-2013.pdf, 2013.

[EVB 96] EVBUOMWAN N.F.O., SIVALOGANATHAN S., JEBB A., "A survey of design philososphies, models, methods and systems", *Journal of Engineering Manufacture*, vol. 210, Part B, pp. 301–320, 1996.

[FAB 12] FABIANI J.-N., *Ces histoires insolites qui ont fait la médecine, vol. 2, les transplantations*, Editions Plon, Paris, 2012.

[FAL 90] FALZON P., BISSERET A., BONNARDEL N. *et al.*, "Les activités de conception: l'approche de l'ergonomie cognitive", *Recherches sur le Design seminar*, Compiègne, 17–19 October 1990.

[FAU 07] FAUCHEUX M., FOREST J., "Expliquer l'inexplicable: Sciences de la conception et créativité", *Cahiers de RECITS*, vol. 5, pp. 211–222, 2007.

[FAU 11] FAUCHEUX M., *Auguste et Louis Lumière*, Gallimard, Paris, 2011.

[FAU 12] FAUCHEUX M., FOREST J., *New Elements of Technology*, UTBM Editions, Belfort, 2012.

[FAU 13] FAUCONNIER P., "Les français inventent beaucoup... au profit de l'étranger", *Le nouvel obs*, available at: http://tempsreel.nouvelobs.com/education/20131011.OBS0868/les-francais-inventent-beaucoup-au-profit-des-etrangers.html, 2013.

[FAY 05] FAYOLLE A., *Introduction à l'entrepreneuriat*, Dunod, Paris, 2005.

[FEL 05] FELDER R., SHEPPARD S., SMITH K., "A new journal for a field in transition", *Journal of Engineering Education*, vol. 94, no. 1, pp. 7–10, 2005.

[FER 82] FERRY J., "Instructions officielles 2 août 1882 : écoles primaires publiques", available at: https://www.samuelhuet.com/fr/paid/41-textes-officiels/941-io-ferry-ecoles-primaires-publiques.html, 1882.

[FLO 02] FLORIDA R., *The Rise of the Creative Class: And How It's Transforming Work, Leisure, Community and Everyday Life*, Basi Books, New York, 2002.

[FON 13] FONTANA R., NUVOLARI A., SHIMIZU H. *et al.*, Reassessing patent propensity: evidence from a data-set of R&D awards 1977–2004, working paper, series number 2013/09, Technical University of Lisbon, March 2013.

[FOR 01] FOREST J., MÉHIER C., "J.R. Commons and H.A. Simon on the concept of Rationality", *Journal of Economic Issues*, vol. XXXV, no. 3, pp. 591–605, 2001.

[FOR 03] FOREST J., MICAELLI J.-P., *Artificialisme: Introduction à une théorie de la conception,* Presses Polytechniques et Universitaires Romandes, Lausanne, 2003.

[FOR 07] FOREST J., MICAËLLI J.-P., Artefact, les apports de l'approche simonienne, MEOTIC seminar, Institut de la Communication et des Médias (Université Stendhal), Grenoble, 7–8 March 2007.

[FOR 11] FOREST J., PATOUILLARD C., "The spread of sustainable urban drainage systems for managing urban stormwater: an MLP analysis", *Journal of Innovation Economics, special edition: Développement Durable*, vol. 2, no. 8, available at: http://www.cairn.info/article.php?ID_ARTICLE=JIE_008_0141, 2011.

[FOR 14] Forest J., "Petite histoire des modèles d'innovation", in BOUTILLIER S., GALLAUD D., FOREST J. *et al.* (eds), *Principes d'économie de l'innovation*, Peter Lang, Bern, 2014.

[FOR 15] FOREST J., HAMDOUCH A., *Quand l'innovation fait la ville durable*, Presses Polytechniques et Universitaires Romandes, Lausanne, 2015.

[FOU 00] FOURNIER M., "Liberté, créativité, autonomie... un siècle d'éducation nouvelle", *Sciences Humaines*, 2000/08-09, pp. 14–15, 2000.

[FRE 13] FRENCH MINISTRY OF EDUCATION, "Le référentiel de compétences des métiers du professorat et de l'éducation, Arrêté du 1er juillet 2013, Journal officiel du 18 juillet", available at: http://www.education.gouv.fr/pid25535/bulletin_officiel.html?cid_bo=73066, 2013.

[FRE 14] FRERY F., "L'innovation ce n'est pas la créativité", *Harvard Business Review France*, available at: http://www.hbrfrance.fr/chroniques-experts/2014/10/4182-linnovation-ce-nest-pas-la-creativite/, 2014.

[GAR 12] GAREL G., MOCK E., *La fabrique de l'innovation*, Dunod, Paris, 2012.

[GAU 84] GAUDIN T., "L'ingénieur et l'innovation", *Culture Technique*, no. 12, pp. 133–135, 1984.

[GOD 09] GODELIER E., "Avez-vous un garage? Le mythe darwinien du créateur", *La tribune*, available at: http://www.latribune.fr/entreprises-finance/tpe-pme/la-tribune-du-manager/articles-non-visibles/20090330trib000361284/avez-vous-un-garage-le-mythe-darwinien-du-createur.html, 2009.

[GOD 10] GODET M., DURANCE P., MOUSLI M., Créativité et innovation dans les territoires, Conseil d'analyse économique, La Documentation française, 2010.

[GOD 14] GODIN B., "Une histoire intellectuelle de l'innovation. De l'interdit politique à la politique publique", in BOUTILLIER S., GALLAUD D., FOREST J. *et al.* (eds), *Principes d'économie de l'innovation*, Peter Lang, Bern, 2014.

[GOM 89] GOMORY R., "From the ladder of Science to the Product Development Cycle", *Harvard Business Review,* available at: https://hbr.org/1989/11/from-the-ladder-of-science-to-the-product-development-cycle, pp. 99–105, 1989.

[GUE 14] GUELLERIN C., "Education: la créativité autant que les savoirs...", *les échos.fr*, available at: http://archives.lesechos.fr/archives/cercle/2014/02/14/cercle_91211.htm, 14 February 2014.

[GUI 50] GUILFORD J.-P., "Creativity", *The American Psychologist*, vol. 5, no. 9, pp. 444–454, 1950.

[HAT 06] HATCHUEL A., LE MASSON P., WEIL B., *Les processus d'innovation conception innovante et croissance des entreprises*, Hermes-Lavoisier, Paris, 2006.

[HOL 09] HOLLANDERS H., CRUYSEN VAN A., Design, Creativity and Innovation: A Scoreboard Approach, Working paper, Maastricht University, available at: http://citeseerx.ist.psu.edu/viewdoc/download?doi=10.1.1.924.793&rep=rep1&type=pdf, 2009.

[HOR 12] HWANG V.W., HOROWITT G., *The Rainforest: The Secret to Building the Next Silicon Valley*, Regenwald, 2012.

[HUY 13] HUYGHE P.-D., Cycle de conférences sur l'Innovation de l'ENSCI, Tuesday 8 October, available at: http://www.ensci.com/actualites/une-actualite/news/detail/News/18248/, 2013.

[IRI 07] IRISH CENTER FOR DESIGN INNOVATION., The Design Difference. A Survey of Design and Innovation amongst Ireland's SME's, Sligo, Ballinode, 2007.

[JAC 11] JACOMY B., "L'innovation technique au fil du temps", *Cahiers du Musée des Confluences*, vol. 7, pp. 59–71, 2011.

[JAC 94] JACOMY B., *Une histoire des techniques*, le Seuil, Paris, 1994.

[JEE 06] JEE, "Special report: the research agenda for the new discipline of engineering education", *Journal of Engineering Education*, vol. 95, no. 4, pp. 259–261, 2006.

[KAP 96] KAPR A., *Johann Guteneberg The Man and his Invention*, Scholar Press, Brookfield, 1996.

[KLI 85] KLINE S., Research, Invention, Innovation and Production: Models and Reality, Report INN-1, Mechanical Engineering Department, Stanford University, March 1985.

[KLI 86] KLINE S., ROSENBERG N., "An overview of innovation", In LANDAU R., ROSENBERG N. (eds), *The Positive Sum Strategy: Harnessing Technology for Economic Growth*, National Academy Press, Washington, 1986.

[KOE 64] KOESTLER A., *The Act of Creation*, Arkana, London, 1964.

[LAB 09] LABREURE D., *Louis Ferdinand Céline: une pensée médicale*, Editions Publibook, Université Saint Denis, 2009.

[LAM 06] LAMARD P., LEQUIN Y-C., *La technologie entre à l'université: Compiègne, Sevenans, Belfort-Montbéliard..., Belfort*, UTBM Editions, Belfort, 2006.

[LAN 08] LANDRY C., WOOD P., "The intercultural city: planning for diversity advantage", Earthscan, London, 2008.

[LAR 11] LAREDO P.H., "L'Europe de la recherche et de l'innovation en transition", in LESOURME J., RANDET D. (eds), *La recherche et l'innovation en France, Futuris 2011*, Editions Odile Jacob, Paris, 2011.

[LED 12] LE DANTEC F., *Contre la Métaphysique, questions de méthode*, Félix Alcan, Paris, 1912.

[LEH 13] LEHN J.-M., "Ne pas éteindre la lumière du futur", *Science et pseudo sciences*, no. 304, pp. 13–21, 2013.

[LEM 99] LE MOIGNE J.-L., *Les épistémologies constructivistes*, Presses Universitaires de France, Paris, 1999.

[LES 04] LESTER R., PIORE M., *Innovation. The Missing Dimension*, Harvard University Press, Harvard, 2004.

[LUB 03] LUBART T., MOUCHIROUD C., TORDJMAN S. *et al.*, *Psychologie de la créativité*, Armand Colin, Paris, 2003.

[LUM 35] LUMIERE A., "Préface", in CELLIER R. (ed.), *Manuel pratique d'illusionnisme et de prestidigitation*, Payot, Paris, 1935.

[LUP 13] LUPIERI S., "La friteuse Actifry: une technologie de rupture", *les échos.fr*, available at: http://www.lesechos.fr/industrie-services/dossiers/0202674284789/0202674285216-la-friteuse-actifry-une-technolgie-de-rupture-554867.php?7whq3SWJ0pcBip Dw.99, 2013.

[MAR 05] MARTINDALE C., "Biological bases of creativity", in R.J. Sternberg (ed.), *Handbook of Creativity*, Cambridge University Press, 2005.

[MAS 01] MASKELL P., "Knowledge creation and diffusion in geographic clusters", *International Journal of Innovation Management*, vol. 5, no. 2, pp. 213–237, 2001.

[MAU 68] MAUNOURY J-L., *La genèse des innovations: la création technique dans l'activité de la firme*, Presses Universitaires de France, Paris, 1968.

[MCW 08] MCWILLIAM E., PRONNICK P., TAYLOR P.-G., "Re-designing science pedagogy: reversing the flight from science", *Journal of Science Education and Technology*, vol. 17, pp. 226–235, 2008.

[MEL 10] MELANDER C., Case study: Danish design support programmes: developing the design skills of Danish businesses, *Design for Europe*, available at: http://designforeurope.eu/case-study/danish-design-support-programmes, 2010.

[MÉN 11] MÉNISSIER T., "Philosophie et innovation, ou philosophie de l'innovation?", *Klesis*, no. 18, pp. 10–27, 2011.

[MES 12] MESR, 15 mesures pour augmenter l'impact économique de la recherche, available at: https://cache.media.enseignementsup-recherche.gouv.fr/file/transfert/05/2/DP15_mesures_pour_le_transfert_de_la_recherche_232052.pdf, 2 November 2012.

[MOR 09] MORAND P., MANCEAU D., Pour une nouvelle vision de l'innovation, Report, Ministère de l'économie de l'industrie et de l'emploi, 2009.

[MUR 10] MURRAY R., CAULIER-GRICE J., MULGAN G., *The Open Book of Social Innovation*, The Young Foundation, London, 2010.

[MUS 98] MUSTAR P. (ed.), *Les chiffres clés de la science et de la technologie*, Economica, Paris, 1998.

[NAT 91] NATIONAL RESEARCH COUNCIL, *Improving Engineering Design: Designing for Competitive Advantage*, National Academy Press, Washington, 1991.

[NOA 13] NOAILLES-SIMÉON P., "De l'entrepreneur à l'innovateur dans une économie dynamique", in BOUTILLER S., DJELLAL F., UZUNIDIS D. (eds), *L'innovation : Analyser, anticiper, agir*, Peter Lang, Bern, 2013.

[NOO 00] NOOTEBOOM B., "Learning by interaction: absorptive capacity, cognitive distance and governance", *Journal of Management and Governance*, no. 4, pp. 69–92, 2000.

[NOO 07] NOOTEBOOM B., HAVERBEKE W., DUYSTERS G. *et al.*, "Optimal cognitive distance and absorptive capacity", *Research Policy*, vol. 36, no. 7, pp. 1016–1034, 2007.

[OEC 05] OECD, *Manuel d'Oslo: Principes directeurs pour le recueil et l'interprétation des données sur l'innovation, 3rd edition*, OECD, 2005.

[OEC 10] OECD, *Measuring Innovation: A New Perspective*, OECD, 2010.

[OEC 15] OECD, "Vers un système d'éducation plus inclusif en France", available at: https://www.oecd.org/fr/france/vers-un-systeme-d-education-plus-inclusif-en-france.pdf, 2015.

[PAH 84] PAHL G., BEITZ W., *Engineering Design: A Systematic Approach*, Springer, London, 1984.

[PAP 01] PAPON P., *L'europe de la science et de la technologie*, Presses universitaires de Grenoble, Grenoble, 2001.

[PHE 13] PHELPS E., *Mass Flourishing: How Grassroots Innovation Created Jobs, Challenge, and Change*, Princeton University Press, New Jersey, 2013.

[PIC 14] PICCARD B., *Changer d'altitude: quelques solutions pour mieux vivre sa vie*, Stock, Paris, 2014.

[PIS 16] PISANI-FERRY J., *Quinze ans de politiques d'innovation en France*, Report, Commission nationale d'évaluation des politiques d'innovation, available at: http://www.strategie.gouv.fr/sites/strategie.gouv.fr/files/atoms/files/fs_rapport_cnepi_21012016_0.pdf, 2016.

[PLA 13] PLATO, *Œuvres complètes*, Arvensa Editions, ePub, December 2013.

[POL 05] POLICASTRO E., GARDNER H., "From case studies to robust generalizations: an approach to the study of creativity", in STERNBERG R.-J. (ed.), *Handbook of Creativity*, Cambridge University Press, Cambridge, 2005.

[PON 03] PONS A., "De l'invention chez Vico", *Congrès international La deuxième Renaissance*, Tokyo, April, 1984, document published online on the Forum du Conseil scientifique MCX-APC, 2003.

[PUE 06] PUECH M., *Homo Sapiens Technologicus*, Editions le Pommier, Paris, 2006.

[REZ 73] REZNIKOFF M.-G., DOMINO C., BRIDGES C. *et al.*, "Creative abilities in identical and fraternal twins", *Behavior Genetics*, vol. 3, no. 4, pp. 365–377, 1973.

[ROB 06] ROBINSON K., "How schools kill creativity", *Ted Conference*, available at: http://www.ted.com, 2006.

[ROG 54] ROGERS C., "Toward a theory of creativity", *ETC: A Review of General Semantics*, vol. 11, no. 4, pp. 249–260, 1954.

[ROY 86] ROY R., WIELD D., *Product Design and Technological Innovation*, Open University Press, Milton Keynes, Philadelphia, 1986.

[SCH 12] SCHUMPETER J., *The Theory of Economic Development: An Inquiry into Profits, Capital, Credit, Interest, and the Business Cycle*, Transaction Publishers, New Brunswick, 1912.

[SCH 15] SCULTE B, "Do these eight things and you will be more creative and insightful, neuroscientists say", The Washington Post, 6 July 2015.

[SCH 39] SCHUMPETER J., *Business Cycles: A Theoretical, Historical and Statistical Analysis of the Capitalist Process*, McGraw Hill, New York and London, 1939.

[SER 15] SERRES M., "Savoir, invention, éducation", *Les cahiers millénaire 3*, no. 31, pp. 47–57, 2015.

[SIG 94] SIGAUT F., "Les points de vue constitutifs d'une science des techniques: essai de tableau comparatif", in PERRIN J. (ed.), *Construire une science des techniques*, L'interdisciplinaire, Limonest, 1994.

[SIM 69] SIMON H.-A., *The Sciences of the Artificial*, MIT Press, Cambridge, 1969.

[SIM 75] SIMONTON D-K., "Sociocultural context of individual creativity: A transhistorical time-series analysis", *Journal of Personality and Social Psychology*, vol. 32, no. 6, pp. 1119–1133, 1975.

[SIM 84] SIMON H.-A., "The structure of ill-structured problems", in CROSS N. (ed.), *Developments in Design Methodology*, John Wiley & Sons, Chichester, 1984.

[SIM 89] SIMONDON G., *Du mode d'existence des objets techniques*, 1st ed. 1958, Editions Aubier, Paris, 1989.

[SIM 92] SIMON H.-A., "Methodological foundations of economics." In AUSPITZ J.L., GASPARSKI W.W., MLICKI M.M. *et al.* (eds), *Praxiologics and the Philosophy of Economics, vol. 1,* Transaction Publishers, New Jersey, 1992.

[SIM 95] SIMON H.-A., "Problem forming, problem finding and problem solving in design", in COLLEN A., GASPARSKI W.W. (eds), *Design and System, Praxiology*, Transaction Publishers, New Jersey, 1995.

[SMI 12] SMILES S., *The Life of George Stephenson, Railway Engineer*, Cambridge Library Collection, Cambridge, 2012.

[SOU 74] SOUPAULT R., "Carrel et la chirurgie cardiovasculaire", *Histoire des sciences médicales*, vol. 8, no. 3, pp. 417–423, 1974.

[SWE 11] SWEDISH INDUSTRIAL DESIGN FOUNDATION, Companies that invest in design are more profitable, available at: http://www.svid.se/en/The-benefit-of-design/Analysis-Companies-that-invest-in-design-are-more-profitable/, 2011.

[TAD 09] TADDEI F., Former des constructeurs de savoirs collaboratifs et créatifs: un défi majeur pour l'éducation du 21ème siècle, Report, OCDE, 2009.

[TAR 93] TARONDEAU J.-C., *Stratégie Industrielle*, Vuibert Gestion, Paris, 1993.

[UKD 05] UK DEPARTMENT OF TRADE AND INDUSTRY, "Creativity, design and business performance", *Economics Paper*, vol. 15, 2005.

[VAN 14] VAN EERSEL P., "La vie intérieure: Cédric Villani", *CLES*, available at: http://www.cles.com/enquetes/article/cedric-villani/page/0/1, 2014.

[VER 93] VERIN H., *La gloire des ingénieurs. L'intelligence technique du XVIème au XVIIème siècle*, Editions Albin Michel, Paris, 1993.

[VIL 12] VILLANI C., *Théorème vivant*, Grasset, Paris, 2012.

[WAT 14] WATRELOT P., "Ecole et innovation: je t'aime moi non plus...", *Chronique innovation*, available at: http://philippe-watrelot.blogspot.fr/2014/03/ecole-et-innovation-je-taime-moi-non.html, 25 March 2014.

Index

L, O, P

S, T, U, V

Other titles from

iSTE

in

Innovation, Entrepreneurship and Management

2017

AÏT-EL-HADJ Smaïl
The Ongoing Technological System
(Smart Innovation Set – Volume 11)

BAUDRY Marc, DUMONT Béatrice
Patents: Prompting or Restricting Innovation?
(Smart Innovation Set – Volume 12)

CHALENÇON Ludivine
Location Strategies and Value Creation of International Mergers and Acquisitions

CHAUVEL Danièle, BORZILLO Stefano
The Innovative Company: An Ill-defined Object
(Innovation Between Risk and Reward Set – Volume 1)

CORSI Patrick
Going Past Limits To Growth

D'ANDRIA Aude, GABARRET Inés
Building 21st Century Entrepreneurship
(Innovation and Technology Set – Volume 2)

DAIDJ Nabyla
Cooperation, Coopetition and Innovation
(Innovation and Technology Set – Volume 3)

GUILHON Bernard
Innovation and Production Ecosystems
(Innovation between Risk and Reward Set – Volume 2)

LALLEMENT Rémi
Intellectual Property and Innovation Protection: New Practices and New Policy Issues
(Innovation between Risk and Reward Set – Volume 3)

LAPERCHE Blandine
Enterprise Knowledge Capital
(Smart Innovation Set – Volume 13)

LEBERT Didier, EL YOUNSI Hafida
International Specialization Dynamics
(Smart Innovation Set – Volume 9)

MAESSCHALCK Marc
Reflexive Governance for Research and Innovative Knowledge
(Responsible Research and Innovation Set – Volume 6)

MASSOTTE Pierre
Ethics in Social Networking and Business 1: Theory, Practice and Current Recommendations
Ethics in Social Networking and Business 2: The Future and Changing Paradigms

MASSOTTE Pierre, CORSI Patrick
Smart Decisions in Complex Systems

MEDINA Mercedes, HERRERO Mónica, URGELLÉS Alicia
Current and Emerging Issues in the Audiovisual Industry
(Diverse and Global Perspectives on Value Creation Set – Volume 1)

MICHAUD Thomas
Innovation, Between Science and Science Fiction
(Smart Innovation Set – Volume 10)

PELLÉ Sophie
Business, Innovation and Responsibility
(Responsible Research and Innovation Set – Volume 7)

SAVIGNAC Emmanuelle
The Gamification of Work: The Use of Games in the Workplace

SUGAHARA Satoshi, DAIDJ Nabyla, USHIO Sumitaka
Value Creation in Management Accounting and Strategic Management: An Integrated Approach
(Diverse and Global Perspectives on Value Creation Set –Volume 2)

UZUNIDIS Dimitri, SAULAIS Pierre
Innovation Engines: Entrepreneurs and Enterprises in a Turbulent World
(Innovation in Engineering and Technology Set – Volume 1)

2016

BARBAROUX Pierre, ATTOUR Amel, SCHENK Eric
Knowledge Management and Innovation
(Smart Innovation Set – Volume 6)

BEN BOUHENI Faten, AMMI Chantal, LEVY Aldo
Banking Governance, Performance And Risk-Taking: Conventional Banks Vs Islamic Banks

BOUTILLIER Sophie, CARRE Denis, LEVRATTO Nadine
Entrepreneurial Ecosystems (Smart Innovation Set – Volume 2)

BOUTILLIER Sophie, UZUNIDIS Dimitri
The Entrepreneur (Smart Innovation Set – Volume 8)

BOUVARD Patricia, SUZANNE Hervé
Collective Intelligence Development in Business

GALLAUD Delphine, LAPERCHE Blandine
Circular Economy, Industrial Ecology and Short Supply Chains
(Smart Innovation Set – Volume 4)

GUERRIER Claudine
Security and Privacy in the Digital Era
(Innovation and Technology Set – Volume 1)

MEGHOUAR HICHAM
Corporate Takeover Targets

MONINO Jean-Louis, SEDKAOUI Soraya
Big Data, Open Data and Data Development
(Smart Innovation Set – Volume 3)

MOREL Laure, LE ROUX Serge
Fab Labs: Innovative User
(Smart Innovation Set – Volume 5)

PICARD Fabienne, TANGUY Corinne
Innovations and Techno-ecological Transition
(Smart Innovation Set – Volume 7)

2015

CASADELLA Vanessa, LIU Zeting, DIMITRI Uzunidis
Innovation Capabilities and Economic Development in Open Economies
(Smart Innovation Set – Volume 1)

CORSI Patrick, MORIN Dominique
Sequencing Apple's DNA

CORSI Patrick, NEAU Erwan
Innovation Capability Maturity Model

FAIVRE-TAVIGNOT Bénédicte
Social Business and Base of the Pyramid

GODÉ Cécile
Team Coordination in Extreme Environments

MAILLARD Pierre
Competitive Quality and Innovation

MASSOTTE Pierre, CORSI Patrick
Operationalizing Sustainability

MASSOTTE Pierre, CORSI Patrick
Sustainability Calling

2014

DUBÉ Jean, LEGROS Diègo
Spatial Econometrics Using Microdata

LESCA Humbert, LESCA Nicolas
Strategic Decisions and Weak Signals

2013

HABART-CORLOSQUET Marine, JANSSEN Jacques, MANCA Raimondo
VaR Methodology for Non-Gaussian Finance

2012

DAL PONT Jean-Pierre
Process Engineering and Industrial Management

MAILLARD Pierre
Competitive Quality Strategies

POMEROL Jean-Charles
Decision-Making and Action

SZYLAR Christian
UCITS Handbook

2011

LESCA Nicolas
Environmental Scanning and Sustainable Development

LESCA Nicolas, LESCA Humbert
Weak Signals for Strategic Intelligence: Anticipation Tool for Managers

MERCIER-LAURENT Eunika
Innovation Ecosystems

2010

SZYLAR Christian
Risk Management under UCITS III/IV

2009

COHEN Corine
Business Intelligence

ZANINETTI Jean-Marc
Sustainable Development in the USA

2008

CORSI Patrick, DULIEU Mike
The Marketing of Technology Intensive Products and Services

DZEVER Sam, JAUSSAUD Jacques, ANDREOSSO Bernadette
Evolving Corporate Structures and Cultures in Asia / Impact of Globalization

2007

AMMI Chantal
Global Consumer Behavior

2006

BOUGHZALA Imed, ERMINE Jean-Louis
Trends in Enterprise Knowledge Management

CORSI Patrick *et al.*
Innovation Engineering: the Power of Intangible Networks

Printed and bound by CPI Group (UK) Ltd, Croydon, CR0 4YY
16/09/2026
14963074-0004